THE SCIENCE OF SLEEP

Wallace B. Mendelson

THE SCIENCE OF SLEEP

WHAT IT IS, WHY WE NEED IT, HOW IT WORKS

IVY PRESS

First published in the UK in 2017 by
Ivy Press
An imprint of The Quarto Group
The Old Brewery, 6 Blundell Street
London N7 9BH, United Kingdom
T (0)20 7700 6700 F (0)20 7700 8066
www.QuartoKnows.com

British Library Cataloguing-in-Publication Data
A catalogue record for this book is available from the British Library

ISBN: 978-1-78240-441-5

This book was conceived, designed, and produced by
Ivy Press
58 West Street, Brighton BN1 2RA, United Kingdom

Publisher Susan Kelly
Creative Director Michael Whitehead
Editorial Director Tom Kitch
Commissioning Editor Kate Shanahan
Senior Project Editor Caroline Earle
Designer Lisa McCormick
Illustrators Louis Mackay, John Woodcock
Picture researcher Kate Duffy
Assistant Editor Jenny Campbell

Printed in China

10 9 8 7 6 5 4 3 2 1

Wallace B. Mendelson, MD is retired Professor of Psychiatry and Clinical Pharmacology at the University of Chicago and has been engaged in sleep research and clinical care for 40 years. He has been the director of the sleep laboratory at the National Institute of Mental Health Intramural Program, the Sleep Disorder Center at the Cleveland Clinic, and the Sleep Research Laboratory at the University of Chicago. Mendelson has authored or co-authored three books on sleep research and sleep disorders medicine and published numerous papers in academic journals. He is a past president of the Sleep Research Society and the recipient of various honors including a special award for excellence in sleep and psychiatry from the National Sleep Foundation in 2010 and the William C. Dement Academic Achievement Award from the American Sleep Disorders Association in 1999.

Dedication

For Miho
The lass who loved a sailor

Disclaimer

Within this book, particularly in Chapters Seven and Eight, there is information about common sleep disorders and treatments. It is not a substitute for medical evaluation or care. If you have difficulties with your sleep or wakefulness, you should consult your doctor for evaluation or for possible referral to a sleep disorders center.

CONTENTS

INTRODUCTION 6

ONE HUMAN SLEEP 8

TWO SLEEP IN ANIMALS 42

THREE THE EFFECTS OF SLEEP DEPRIVATION 52

FOUR CIRCADIAN RHYTHMS AND SLEEP 72

FIVE A LIFETIME OF SLEEP 86

SIX HORMONES AND SLEEP 98

SEVEN SLEEP DISORDERS 108

EIGHT INSOMNIA 144

GLOSSARY 170
SELECTED BIBLIOGRAPHY 172
INDEX 174
ACKNOWLEDGMENTS 176

INTRODUCTION

Sleep means different things to different people, and indeed its meaning differs in the same person at various times. I remember as a boy going to bed on Christmas Eve, excitedly anticipating the presents awaiting me in the morning, and thinking that since I would soon be asleep, the time would seem to pass in an instant. Then the presents would be mine. Not surprisingly, that sort of thinking led instead to a long period of unhappy wakefulness. The opposite can occur as well, as in the song "The Green Green Grass of Home," in which sleep is a time of escape into happy memories, in contrast to the very unfortunate events awaiting the sleeping prisoner in the morning. For others, sleep can become a kind of testing ground: a person who prides herself on always being the best at whatever she does can view good sleep as a challenge, something she has to work at—the result, paradoxically, being poor sleep. It can also be a time of anxiety. People for whom it is important to feel in control of things can find it worrisome to have a period each night in which they seem more vulnerable and not in charge. For others, sleep can be a time of getting a glimpse of the "real" world; I have had patients who say that their dream experiences during sleep seem so much more real and meaningful than what they awaken to in the morning.

Sleep is inextricably tied to a world of alternating day and night, and our bodies have developed elaborate mechanisms to time our waking and sleep accordingly. As we will discuss later, no one is certain what this entails physiologically—it is not just a simple matter of increasing metabolic energy stores—but its presence (or absence) plays a role in what we think about sleep. Related to this is the notion of sleep as a pleasurable experience, something to look forward to. Sadly, for many people the opposite is true. The genesis of this is not always clear. Some think that a lifelong feeling that sleep is an unhappy time is a derivative of childhood experiences, in which the more typical learning association of sleeping with pleasure did not take place. Others view this as a disorder of the amount of brain chemicals that normally bring on sleep. Another view is that it may result from habits in which bedtime is used for behaviors incompatible with sleep, such as worrying and planning tomorrow's battles.

Sleep is also inextricably tied to the environment in which we live, in a world of alternating day and night. Our bodies have developed elaborate mechanisms to help time our waking and sleep to be in conjunction with light and darkness. Sometimes this timing can go astray, either due to behaviors such as engaging in shift work or flying long distances, or due to inherent problems of the body clock. These in turn can lead to difficulties with sleeping, or at least with sleeping during the traditional hours allocated for it.

Sleep can also be a kind of social behavior, inside the species, or across species (for instance when sleeping with a pet). We often use the euphemism of sleeping together to refer to another kind of activity that can take place in bed, but this kind of delicate phraseology can obscure another aspect, which is that repetitive sharing of the sleep experience may play a role in a couple bonding together.

There is also a sense that sleep is important to health, both physical and mental. Sleep which is curtailed or disrupted can lead, for instance, to a predilection to diabetes and related disorders. It seems to be important for the formation of long-term memories. This suggests, for instance, the futility of students doing "all nighters" of studying. It turns out that getting a good night's sleep may be the most helpful thing in preparing for an exam in the morning. One of the great believers in a good night's sleep, incidentally, was Alexander the Great. In 331 BCE, before the critical battle of Gaugamela in which he overwhelmed a vastly larger Persian army on their own territory, he slept so deeply that his officers became worried and had to awaken him. He got up, put on his armor, and went on to an outstanding victory which set the stage for conquering an empire.

In this book we will present the scientific understanding of sleep, beginning by describing its basic processes and how to measure them. It will be seen that sleep results from the careful orchestration of a variety of physiologic processes. As in any complex mechanism, sometimes things go awry, in this case resulting in clinical sleep disorders which are experienced as insomnia, excessive sleepiness or undesirable behaviors during sleep. We will describe some of these disorders, and some of the treatments that are available. This information is not a substitute for medical evaluation. If you think you may have a sleep disorder, you should consult your doctor for evaluation and possible referral to a sleep disorders center. It is hoped that, armed with the information provided here, you will be better able to understand and discuss what is happening, and to make more informed choices in conjunction with your doctor.

Just as sleep is a universal human behavior, so is human curiosity and the desire to know more about ourselves. A number of men and women devoted themselves to learning more about sleep, long before sleep studies became an established scientific discipline. They came from a variety of unlikely backgrounds—a World War I cavalryman and a fighter pilot, for example. One was looking for something entirely else, the basis of a supposed "psychic energy" which might let people communicate across long distances, and ended instead with the groundbreaking discovery of the human electroencephalogram. Another was a doctor faced with treating patients in a worldwide epidemic of encephalitis, who recognized a pattern to the parts of the affected brains—and learned from it the basic structures making it possible for us to be awake or asleep. Another had made his fame developing a method to measure the speed of projectiles for the Army, but curiosity led him to measurements of many other kinds of things, including electrical waveforms during the human sleep stages. If you, the reader, have picked up this book, it sounds like you, too, have curiosity about how things work, and it is my hope that here you will learn more about how we sleep and wake.

Chapter One

HUMAN SLEEP

We can measure our lives on many different levels. A life, for instance, could be considered the period in which a person's heart beats a certain number of times (perhaps 3 billion), or enjoys Saturday nights (about 3,600), or experiences periods of peacefulness (all too few), or feels the joy of love (hopefully at least once). In the same way, sleep, which occupies perhaps one-third of our existence, can be viewed on a number of different levels. It is a process that can be measured physiologically while at the same time understood as a psychological experience, and even a social behavior. In this book we will explore some of these aspects of sleep. Human sleep is a reversible period of decreased consciousness and responsiveness, comprised of two distinct states known as rapid eye movement (REM) and non-rapid eye movement (NREM) sleep. Although the modern era of sleep research began in the 1950s with the description of REM sleep, its roots go back to the 1920s with the discovery of the human electroencephalogram, or EEG. Our growing understanding of sleep has been influenced by some remarkable individuals who (sometimes while looking for something else) made crucial observations about sleep, by developments in psychology and technology, and even by world events.

WHAT IS SLEEP?

The many qualities of sleep

Sleep is a period of recurring behavioral quiescence, which has several qualities, including decreased awareness of and responsiveness to the environment, diminished consciousness, and the rhythmic appearance of certain physiologic patterns (stages). It tends to occur at particular parts of the 24-hour day–night cycle, and at customary locations, both depending on the particular species and environment. It is reversible, distinguishing it from coma or ongoing anesthesia. It is also self-regulating: if one is deprived of sleep, there will be a drive to have increased "recovery sleep" to make up for the loss. It is necessary for life, and is present in all mammals. These qualities will be discussed later in the chapter.

During sleep one becomes less aware of the surroundings; this is one of the qualities that distinguishes sleep from quiet wakefulness. An extreme case would be someone who falls asleep at the wheel, consequently running a red light. On the other hand, this process is not absolute; it is clear that we can process, and act on, sensory information

DIMINISHED CONSCIOUSNESS

The nature of diminished consciousness in sleep is by far the hardest to express, since consciousness itself is so poorly understood. When we speak of consciousness, we refer to our experience of self and the world. Perhaps even more fundamentally, consciousness can be defined as "our mode of access." We can additionally speak of "acts" of consciousness, such as perceiving, willing, imagining, and the "contents" of these acts, such as perceiving a sunset, willing a meeting, or imagining a good outcome.

Other efforts to characterize consciousness are helpful. For example, consciousness has been described as a paradoxical state in which a person is simultaneously a subject who can experience things and an object perceived by oneself. Generally most of us recognize this when we speak of ourselves in both ways simultaneously, for instance when we say, "Sometimes I have to remind myself that..." or "I owe it to myself to... ". In such phrases we recognize that we are both a being having experiences and at the same time we can picture ourselves as objects.

Additional qualities of consciousness include consciousness as subjective and private (not available to another person); unitary (experienced by a single person); and characterized by a subjective "feel" of each experience, a "what it is like" aspect. The American philosopher Thomas Nagel (1937–) illustrated the "feel" which is essential to consciousness in the following manner: although we may study and understand the neurophysiology of a bat, we can never know how the world is experienced by a bat. Some authors argue that a subjective experience such as consciousness cannot profitably be studied by traditional scientific techniques; others believe that the fact that a phenomenon is experienced subjectively does not mean that it cannot be explored objectively.

New developments in technology, including the use of brain-imaging studies, are beginning to advance our understanding of the physiology behind consciousness, and researchers have developed a number of models of how it may occur. In addition to looking at normal sleep, neuroimaging studies of people who have been given hallucinogens such as psilocybin have contributed new theories of consciousness based on the notion of entropy—the degree of order and disorder in neural connections.

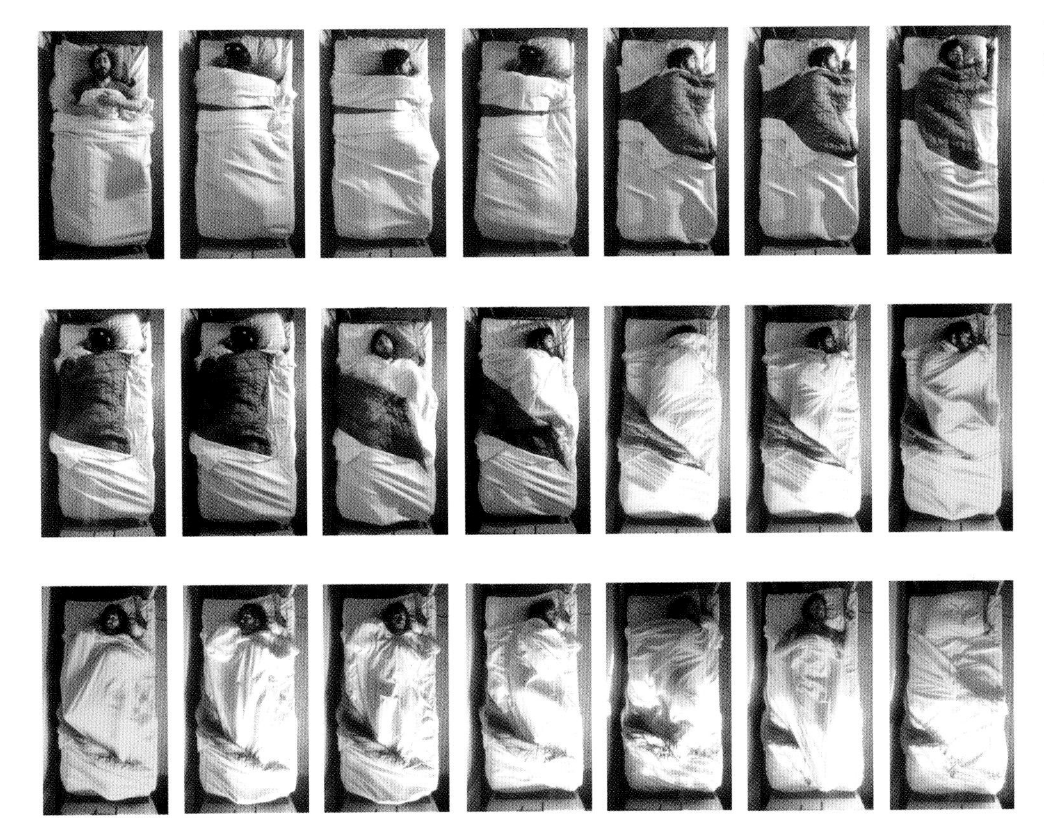

SLEEP STUDIES
This composite picture shows a subject undergoing a sleep recording across the night. The scientific study of sleep continues to reveal more and more insights into what happens while we are asleep, and how it influences our waking lives.

during sleep. A new parent, for instance, can sleep through the noise of a truck driving by the house, but awaken quickly at the sound of the newborn baby crying. Similarly, in a laboratory situation, the volume of a sound required to wake a person up (the "auditory arousal threshold") is much higher for a meaningless sound such as an electronic tone compared to a meaningful stimulus such as a phone ringing or hearing one's name called. A sleeper, then, is able to process information about an incoming sound, and determine whether it is important or not. Again, the arousal response is dependent on a variety of factors including the sleep stage, the duration of wakefulness before sleep, and how far into sleep one is when the sound occurs. Some individuals are more likely to be awakened by low volume noises, and hence are considered to be "light sleepers." Interestingly, people with insomnia often have a normal auditory arousal threshold, suggesting that insomnia is different from just light sleep. Some sleeping pills increase the auditory arousal threshold, raising the possibility that the medicated sleeper will not be aroused by, for instance, a smoke alarm.

Time-lapse videos show that two sleepers adjust their positions in response to the other's movements (whether the other sleeper is a pet or another human). In one film, for instance, a man sleeping with a cat on the bed may turn on his side, with the cat then responding by settling comfortably in the warm nook behind his knees. In co-sleeping humans, an elbow in the ribs may result in an adjustment of the second person's position. Thus, although sensory input is diminished during sleep, this is not absolute, and indeed during sleep we are able to take in and act on information to some degree.

HOW IS THE ELECTRICAL ACTIVITY OF THE BRAIN MEASURED?

Introduction to the Electroencephalogram (EEG)

Sleep is comprised of rhythmically recurring sleep stages, which are defined by looking at brain waves, or an electroencephalogram (EEG), in conjunction with the electrooculogram (EOG), which records eye movements and the electromyogram (EMG), a measure of muscle activity.

The discovery of brain waves

Before the sleep stages are described, it is useful to look at how the brain waves were discovered, and what they are like. It was discovered in the 1820s that electrical current could cause the needle of a magnetic compass to fluctuate, and that this effect could be multiplied by the use of coils of wire. An instrument based on this observation was known as a galvanometer, named after Luigi Galvani (1737–98), an Italian physician and biologist who in 1791 observed that electrical currents could cause the limbs of a dead frog to twitch. This was one of the first observations that electrical currents might be involved in biological processes. Some years later, Richard Caton (1842–1926), a Scottish physiologist, used a galvanometer to detect electrical current from the brains of dogs and apes. In 1875 he reported that the current differed at various times, increasing in strength during sleep and preceding death, then disappearing upon death.

A half century later, Hans Berger (1873–1941), a German psychiatrist, made the next great advance. Berger had been a mathematics student who dropped out and enlisted in the cavalry. One day, he was thrown from his horse, landing in the path of an oncoming cannon carriage, which barely managed to stop at the last moment. At the same time that Berger had this life-threatening experience, his sister living some distance away is said to have had a sudden sensation that he was in danger, and urged her father to contact him. Berger was so struck by this apparent "psychic energy" alarming his sister, that he went on to devote his life to exploring the brain and how objective measures of its activity might relate to subjective psychic processes. In 1929, he described recordings of the electrical waves in humans that had been reported by Caton in animals, and developed a way to record them on moving strips of paper. He coined the term "electroencephalogram" to describe his new discovery. He showed that the waves differed in waking, in sleep, and in anesthesia, and that sharp spiking patterns appeared during epileptic seizures. He described the alpha rhythm in a subject resting with closed eyes, and its disappearance and replacement by faster beta waves when the eyes were opened. These waveforms over the years have been divided into several wavebands according to their frequency (number of waves per second). Other information used in describing them involves their shape, amplitude (a measure of their energy), and location on the head (see also "Features of an oscillating system," page 75). Subsequent to Berger's work focusing on alpha and beta, a series of EEG bands have been further defined and are generally recognized to this day.

MAJOR EEG BANDS

Delta waves (0.5–4 Hz)

These slow waves have a frequency range of 0.5–4 cycles per second (known as cps or Hz). Our interest in them from the point of view of sleep is that they are characteristic of stage N3—also known as slow-wave sleep (SWS), or stages 3 and 4. In clinical EEG work in waking patients, in distinction from sleep studies, they can be associated with the presence of lesions in local areas of the brain, or can appear in a widespread manner in diffuse disorders.

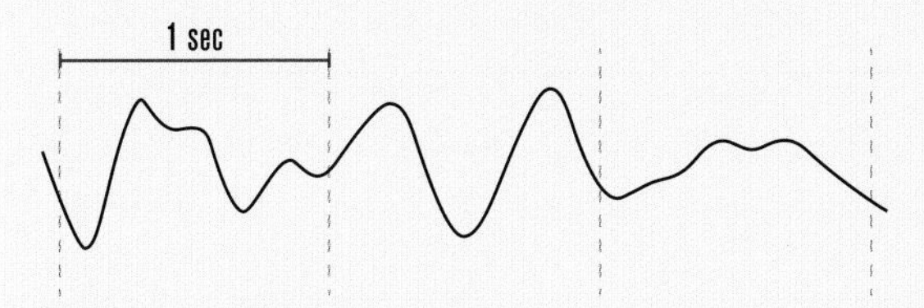

Theta waves (4.5–8 Hz)

These waves appear during lighter sleep. In waking clinical EEG work they are considered a sign of sleepiness, increasing with the duration of wakefulness. They can also appear during hyperventilation. In sleep studies, they are an important rhythm in stage N2 (stage 2) sleep.

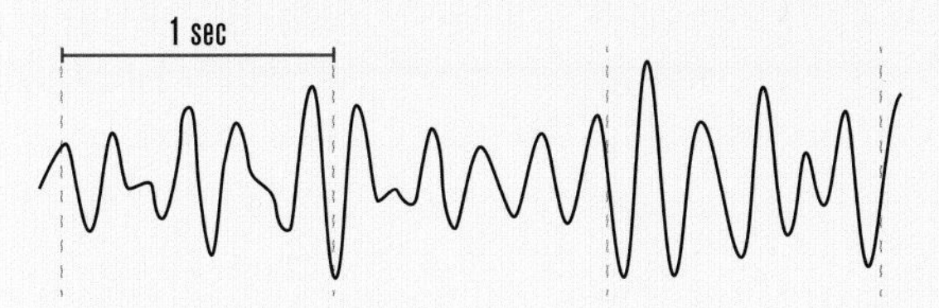

Alpha waves (8.5–12 Hz)

Also known as "Berger waves," these are best seen occipitally (in the back of the head), and are manifested in a relaxed but awake person with eyes closed. As mentioned before, they greatly decrease when the eyes are opened. When eyes are closed in an awake person, they decrease as one becomes sleepy.

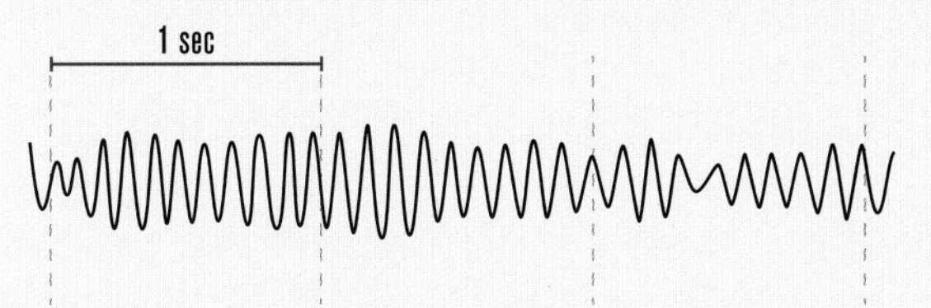

Beta waves (12.5–30 Hz)

Beta waves, which are more evident anteriorly (toward the front of the head), are often divided into irregular, disorganized, low amplitude and organized waveforms. The former is primarily seen in an awake person who is actively thinking, concentrating, or feeling anxiety. The latter is seen in some illnesses or as a result of some sedatives including barbiturates or Valium-like drugs (benzodiazepines). All EEG bands (including beta) can be viewed visually or measured electronically. The amount of electronically measured beta activity is sometimes considered a measure of arousal of the cerebral cortex.

1 sec

THE SLEEP STAGES

Initial discoveries

The next step in the evolution of understanding sleep was the recognition that EEG brainwaves, when combined with other kinds of physiological information, could be used to identify rhythmically recurring, discrete sleep stages. The first description of these sleep stages was made by the American scientist Alfred Loomis (1887–1975). As a young man in the army he developed the Aberdeen Chronograph, a system for measuring projectile velocity by firing a bullet through revolving paper-covered aluminum disks. He went on to a successful career in investment banking. Becoming restless once again, and still remembering his success with the chronograph, he turned his attention to developing radar for military purposes and for ground control during the landing approach of aircraft. He was fascinated with the measuring of waveforms and, among the many projects at his laboratory at Tuxedo Park, New York, was the study of sleep. Using a large 8 foot (2.4 m) diameter recording drum, he described in 1937 a series of five discrete recurring sleep stages during the night, which he rather unpoetically designated as stages A–E. In terms of later development, stages A and B correspond roughly to what was later called stage 1, C corresponded to stage 2, and D and E resembled slow-wave sleep. All together they correspond to what we now call non-rapid eye movement (NREM) sleep.

The discovery of rapid eye movement (REM) sleep

The next big development, which in effect ushered in the modern age of sleep research, occurred in the early 1950s. Nathaniel Kleitman (1895–1999), a physiologist at the University of Chicago, had been interested in eye movements and blinking as a marker of sleep onset and depth of sleep, as well as possible rhythmic behaviors, in infants. He enlisted the aid of a graduate student, Eugene Aserinsky (1921–98). Following observations in infants, they adapted the technique of the electrooculogram (EOG) for continuous use in sleep of children and adults. In the process of doing so, they observed the periodic appearance of vigorous and jerky ocular activity. This new stage, known as rapid eye movement (REM) sleep, was characterized not only by the eponymous eye movements, but also by relaxation of the major weight-bearing muscles, irregularity of respiratory and heart rate, and loss of temperature control. It also has psychological counterparts, and most dreaming, in the conventional sense of the word, occurs in REM. Indeed, REM sleep is as different

RAPID EYE MOVEMENTS
An example of EOG (electrooculogram) channels in a polysomnogram showing the rapid eye conjugate eye movements that occur during REM sleep. The discovery of REM sleep represented an exciting breakthrough in sleep research.

ALFRED LOOMIS
Alfred Lee Loomis (1887–1975) made contributions in a wide area including ballistics, radio-based navigation, radar, and ground control of aircraft. In the sleep field he characterized the K-complex and created the first major classification of the sleep stages.

NATHANIEL KLEITMAN
Nathaniel Kleitman (1895–1999), the co-discoverer of REM sleep, left a legacy of research in a wide variety of areas including the sleep–wake cycle, sleep deprivation, and circadian rhythms. Among his studies were recordings of sleep in settings in which the normal cues of daytime and night-time were not evident, so that he could manipulate waking and sleeping schedules at will. He is pictured here in 1938, studying a subject in Mammoth Cave, Kentucky, to see the physiologic consequences of living on a 28-hour sleep–wake cycle.

from the rest of sleep (dubbed by Kleitman as non-REM or NREM sleep) as NREM is from waking. This has led some authors to describe humans as having three distinct states of consciousness: waking, REM, and NREM sleep.

The sleep stages do not appear randomly, but instead are manifest in a rhythmic, repetitive pattern throughout the night. There are a number of influences on the appearance and duration of the individual stages, including a basic approximately 90–100-minute innate rhythmic cycle of NREM and REM sleep (an example of "ultradian rhythms," see page 75), the time of the 24-hour day at which sleep occurs, and the duration of wakefulness before sleep. In the next section we will describe the sleep stages in much more detail, but perhaps the important message at this point is that sleep is not a unitary process, but is comprised of two very distinct states as well as several distinct stages in NREM sleep.

HOW IS SLEEP ORGANIZED INTO STAGES?

The evolution of sleep staging

The first major revision of sleep staging after Loomis was developed in 1968 by the American scientists Alan Rechtschaffen (1928–) and Anthony Kales (1934–); their *Manual of Standardized Terminology, Techniques and Scoring System for Sleep Stages of Human Subjects* was the standard classification until 2007, when the American Academy of Sleep Medicine (AASM) made revisions which are in effect today and used internationally. The Rechtschaffen–Kales criteria were largely oriented to the use of polygraphs recording with ink pens on paper. Since a page of the typical EEG fan-fold paper was 30 cm wide, ultimately the most common recording speed was with the paper moving 10 mm/sec, so that a 30-second view of sleep was seen on each page. Thus sleep was scored in 30-second "epochs," that is, for each 30-second page, a decision was made of what predominant sleep stage was present. Since a box of paper was typically 1,000 pages long, one night's sleep could be recorded on one box. The signal moving across the paper could be magnified to the degree desired. EEGs are very weak signals measured in millionths of a volt (microvolts or uV); for comparison, electrocardiogram signals from the heart as detected on the skin are roughly one thousand times stronger and are measured in thousandths of a volt (millivolts). Incidentally, the 30-second sleep stage scoring epochs are often used to this day even with modern electronic equipment: it is a good example of how past solutions are carried forward even when new technologies are developed. It's the same reason that now even in the digital age, most popular songs still last about 3 minutes (the duration of recording time that was available on old 45 RPM vinyl discs).

The waking and NREM sleep stages and REM sleep, as classified by the currently used American Academy of Sleep Medicine standard, are as follows:

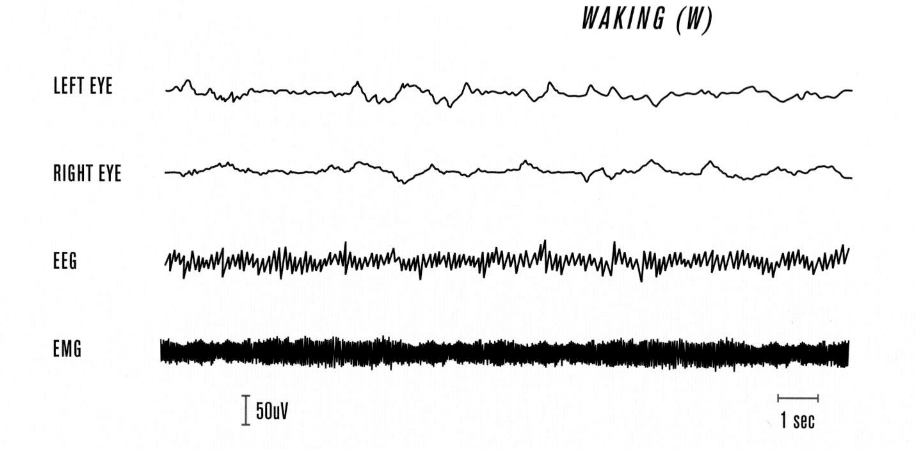

Waking (W)
The EEG shows at least 50 percent alpha activity in a relaxed awake subject whose eyes are closed. There is no significant eye movement activity and there is good muscle tone from the submental electromyogram (EMG—see page 22).

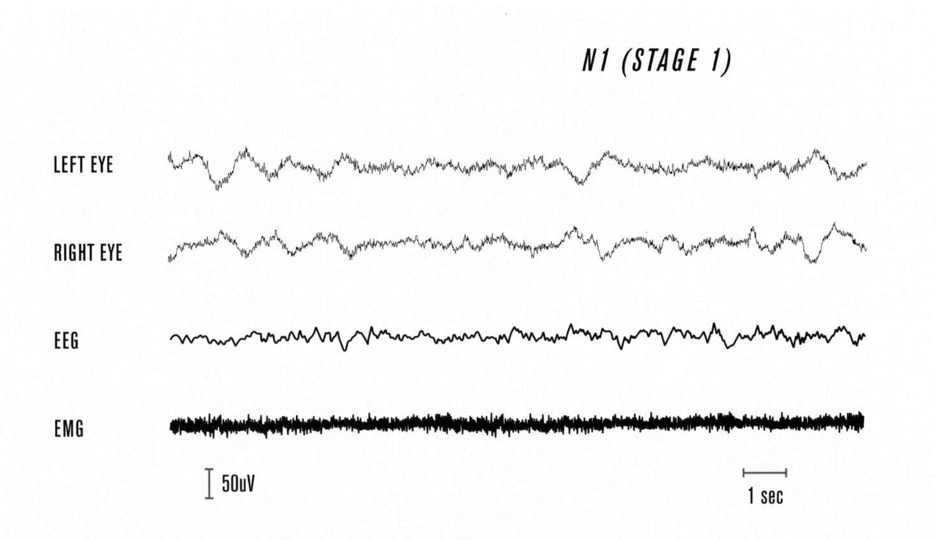

N1 (stage 1)

As a person enters this transitional stage, alpha activity drops to less than 50 percent of the epoch, and the remainder is comprised of low voltage, mixed frequency signals. In some people, alpha activity during waking is poorly defined, so that its decline is hard to recognize. There may be sharp transient EEG waves at the vertex (the top of the head) in the central midline area. There is an absence of jerky saccadic eye movements (rapid step-like movements which occur when fixing on an object), although there may be slow rolling eye movements as seen above. There is an absence of EEG spindles and K complexes, which we will talk about later, as part of N2 sleep.

The behavioral and psychological manifestations of sleep appear in N1 and early N2 sleep. If a person is given a simple task such as tapping two keys alternately, (s)he will generally cease to do this after a few seconds of N1. Similar decreases in perception of visual or auditory stimuli occur. The subjective experience of perceiving oneself as awake gradually disappears. In one study, subjects were aroused at different intervals after the beginning of sleep and asked what they experienced; it was only after several minutes of EEG-defined sleep that most reported that they thought that they had been asleep. The transition to sleep can also be marked by brief jerking movements of the body. These are known as hypnic myoclonia, and are a normal phenomenon, though they may occur more often during periods of stress or changing sleep schedules.

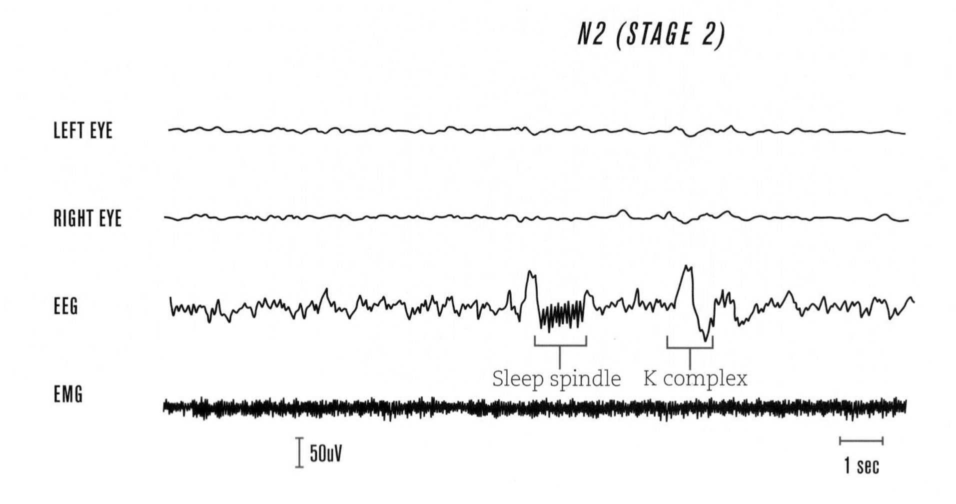

N2 (stage 2)

The background EEG activity is of mixed frequency and low voltage, but now two kinds of transient waveforms appear. The first of these, sleep spindles, are brief episodes of rhythmic activity at a frequency of 12–14 Hz, lasting at least 0.5 sec. In addition there are K complexes, slower high-voltage negative waves followed by positive waves, sometimes with superimposed 12–14 Hz spindle activity. The slow, rolling eye movements of N1 are no longer present. Sometimes the slow delta EEG waves characteristic of N3 are present, but if so occupy less than 20 percent of the epoch.

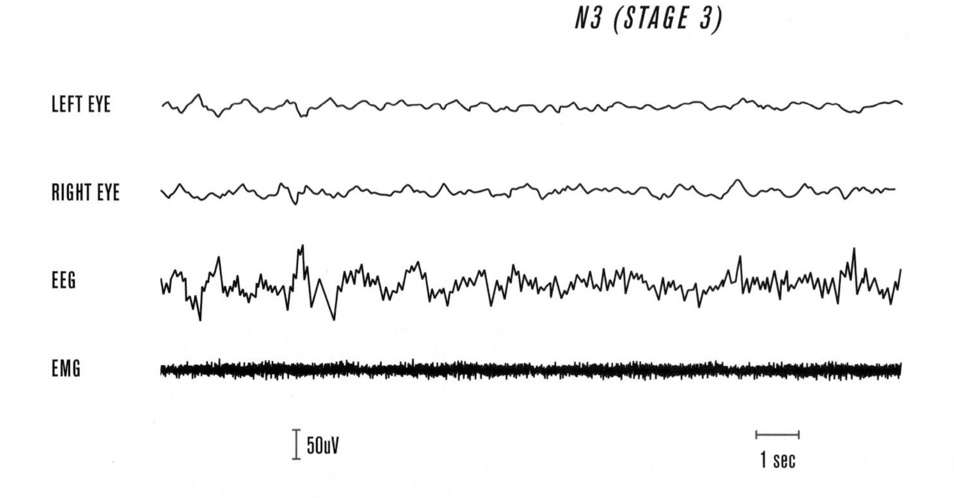

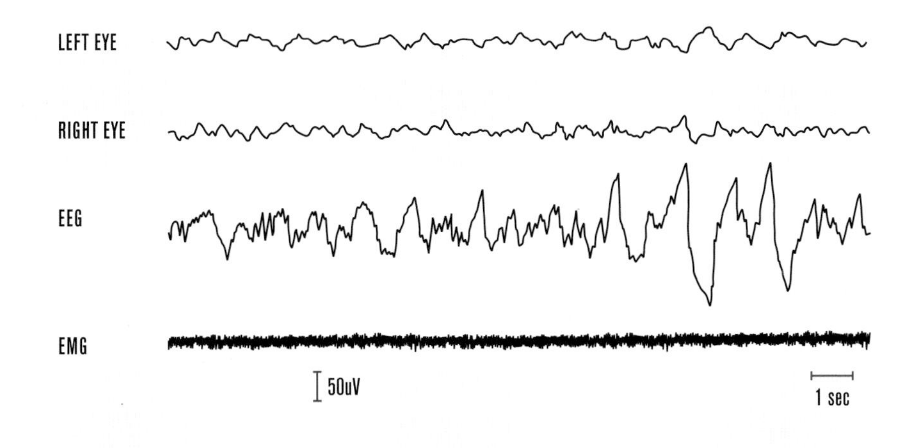

N3 (stages 3 and 4)

N3 (also known as slow-wave sleep or delta sleep) is characterized by higher amplitude (greater than 75 uV), slow (0.5–4 Hz) delta waves, which comprise more than 20 percent of the epoch. In the Rechtschaffen–Kales criteria, epochs containing 20–50 percent delta activity were known as stage 3, while those with greater than 50 percent delta were known as stage 4. In the newer AASM criteria, they are combined into one stage, N3.

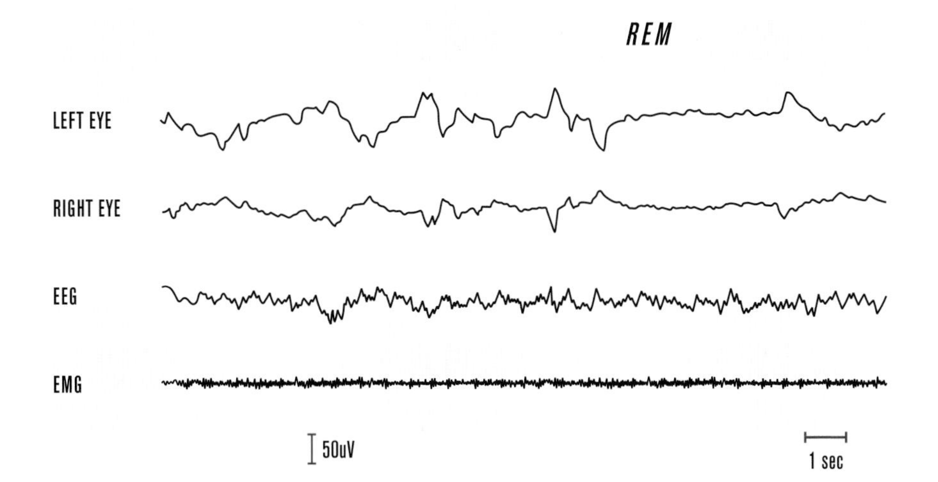

REM sleep

In REM, the EEG is comprised of mixed frequency waves similar to N1. Unlike N2, sleep spindles and K-complexes do not appear. The EMG shows a significant loss of muscle tone. At times, the electrooculogram channel (EOG–see page 22) shows rapid conjugate eye movements.

THE REM–NREM CYCLE

Sleep patterns through the night

The sleep stages appear in a predictable sequence, which repeats about every 90–100 minutes. In a typical night in a young adult, after lights are turned out there is a period of wakefulness before sleep appears (the duration of this period of wakefulness is known as "sleep latency"). Then one enters the NREM sleep stages N1, N2, N3, and, after roughly 90 minutes, REM sleep. (The duration from sleep onset until the beginning of REM is known as "REM latency".) The REM–NREM cycle goes on throughout the night, and a typical night's sleep might contain four to six such cycles. There are different ways of measuring a cycle, but a typical way is from the beginning of one REM period to the beginning of the next.

Influences on the sleep stages

The content of the REM–NREM cycle is altered across the night. The first NREM period is relatively rich in slow-wave (N3) sleep, which diminishes in later cycles. The amount of slow-wave sleep is influenced by the amount of prior wakefulness before sleep—the longer the prior wakefulness, the greater quantity of subsequent slow-wave sleep. For this reason, naps in the afternoon may have more slow-wave sleep than morning naps (since in the afternoon it is longer since the last sleep period). Similarly the amount of slow-wave sleep at night following a daytime nap will be diminished, since it has been a relatively short time since the last sleep.

In contrast to slow-wave sleep, the first REM period of the night is relatively short, and subsequent periods typically become progressively longer. This progression is less clear in the elderly and in some conditions, including depression. The amount of slow-wave sleep is less affected by the timing of sleep. However, in very prolonged normal sleep, or in temporal isolation (experimental situations in which a person lives without clocks or time cues), slow-wave tends to appear roughly every 12 hours, which suggests that there is some influence of basic body rhythms as well.

Unlike slow-wave sleep, the amount of REM is affected more significantly by the time during the 24-hour day. Thus the amount of REM sleep in morning naps is greater than in afternoon or evening sleep, and is relatively less affected by duration of wakefulness before sleep or amount of prior sleep. Age is also an important determinant of the amount of the sleep stages in each sleep cycle (see Chapter Five, pages 86–97). For instance, the sleep of infants and children is relatively rich in slow-wave sleep, which declines across the lifetime and is greatly diminished in the elderly.

PROGRESSION OF SLEEP ACROSS THE NIGHT
The stages of sleep across the night, in a typical young adult. Note the cycles of REM and NREM sleep, lasting about 90 minutes. The first REM period is relatively short, getting longer and longer as the night progresses. Two terms of interest are sleep latency (the time from when lights are turned out until the beginning of sleep), and REM latency (the time from sleep onset until the beginning of the first REM period, is typically 90–110 minutes).

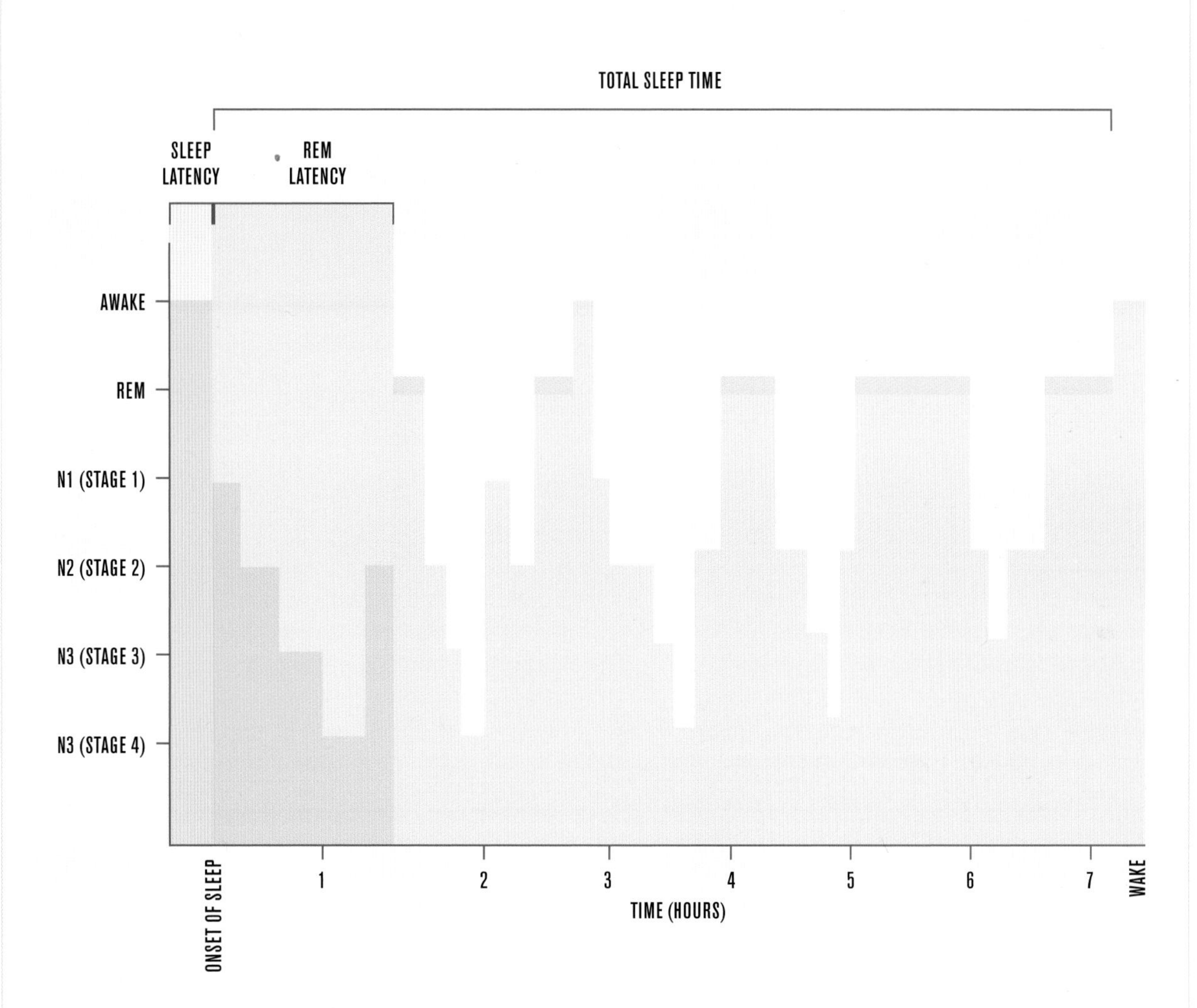
SLEEP STAGES IN A TYPICAL YOUNG ADULT
TOTAL SLEEP TIME
SLEEP LATENCY
REM LATENCY
AWAKE
REM
N1 (STAGE 1)
N2 (STAGE 2)
N3 (STAGE 3)
N3 (STAGE 4)
ONSET OF SLEEP
1
2
3
4
5
6
7
WAKE
TIME (HOURS)

HOW ARE SLEEP STUDIES PERFORMED?

Recording sleep

Sleep staging is determined by recording and combining three sorts of information: the EEG, the electromyogram or EMG (a measure of muscle activity), and the electrooculogram (EOG), which measures eye movement activity. Modern clinical sleep studies, which assess a wide range of other physiologic processes, are done using polysomnography (PSG), and these additional measures such as the electrocardiogram, ventilatory chest movements, and blood oxygen saturation (see Chapter Seven, pages 112–13), are used in assessing clinical sleep disorders.

EEG

This is recorded from electrodes, small metal cups about the size of a dime, placed centrally on the scalp—in its simplest form—with one anteriorly and one posteriorly. This is often done using what is known as unipolar recording, which is the comparison of electrical activity between an electrically active spot (the scalp) and a relatively inactive area (an electrode over the mastoid bone behind and below the ears). Sometimes bipolar recordings, which are more typically used in the waking clinical EEG, are performed, by comparing the electrical currents in two electrically active EEG sites. In either case, the principle is the same: the difference between the electrical activity in two locations is magnified, filtered to select the frequency range desired (the range for EEG in sleep is generally 0–30 Hz), and displayed. From the 1950s for the next several decades, this was typically done on a polygraph machine, which employed galvanometers that might have been recognizable to the earlier nineteenth-century researchers. The galvanometer arm was an ink pen that left lines on a paper tracing. Subsequently, these "ink slingers" have been replaced by devices in which the signal is digitized, recorded, and displayed on a computer screen.

EMG

The electromyogram is used to measure muscle tenseness. Typically, the recording might be between two electrodes placed under the chin (submentally). The resulting display, at a much higher frequency than the EEG, is such that the thicker the EMG line, the higher the muscle tone. This is useful in recording the decreased muscle activity characteristic of REM sleep. In clinical polysomnography, similar pairs of EMG electrodes can be placed on the legs (over the anterior tibialis muscle) to record the jerking leg movements characteristic of periodic leg movement disorder (see Chapter Seven, page 113).

EOG

The electrooculogram is recorded from electrodes on the facial skin near the outer canthi (junction) of the eyelids. As the front of the eye is more electrically positive than the back, when the eyes move laterally, the electrodes record the movement. This is used to detect the slow rolling eye movements sometimes seen in N1 (stage 1) sleep, and more importantly the conjugate and jerky rapid eye movements which gave REM sleep its name.

HOW SLEEP IS MEASURED

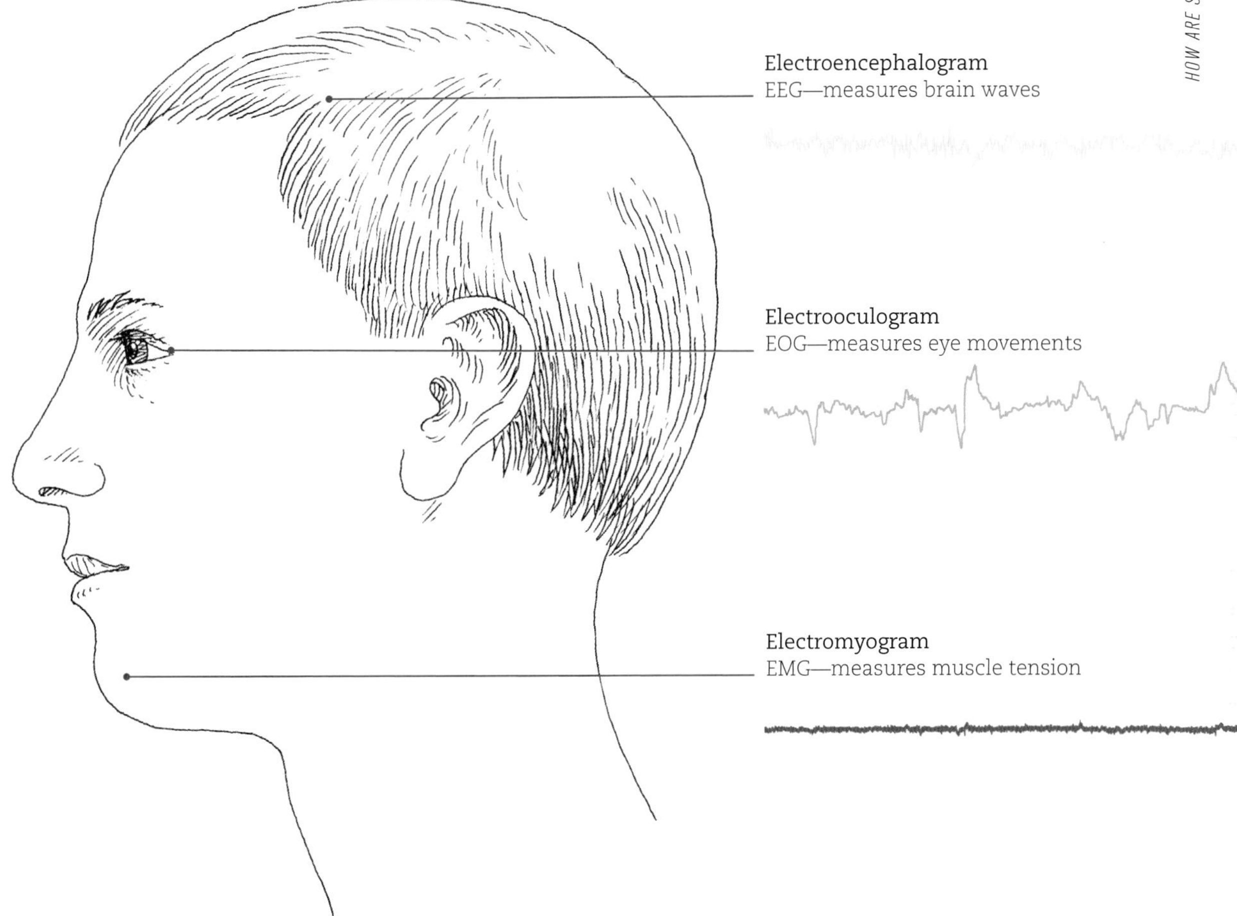

SLEEP STAGING

Sleep recordings involve a wide variety of physiologic measures. The most basic aspect of the recordings involves the information used to determine the stages of sleep. These include the electroencephalogram, electrooculogram, and electromyogram.

WHAT CHANGES OCCUR IN THE BODY IN NREM SLEEP?

The physiology of NREM sleep

A number of aspects of physiology are altered as one enters NREM sleep. The EMG becomes steady and slightly decreased compared to waking, though overall muscle tone remains significant. A number of changes happen in the respiratory system. There is less drive from the brain to the diaphragm and chest wall respiratory musculature (the "respiratory pump muscles"), as well as to the muscles which keep the upper airway open, resulting in increased resistance to airflow though the upper airway. Respiration may transiently become slightly irregular after sleep onset, a phenomenon known as "periodic breathing." There is a decrease in total "minute ventilation" particularly in N3 (slow-wave) sleep. This is largely due to a decline in the total amount of air volume of each breath, rather than in rate of respiration.

Overall, the functioning of the autonomic nervous system is largely preserved in NREM sleep, with dominance of the parasympathetic nervous system relative to the sympathetic nervous system. The function of the heart is largely unchanged. There may be a small coupling of the heart rate with respiration, such that the heart speeds up slightly during inspiration and slows slightly during expiration. This is normal, and indeed the absence of this respiratory coupling may be related to illness and advancing age. Cerebral blood flow and metabolic activity in the brain are mildly decreased in N2 (stage 2) and significantly decline in N3 (slow-wave) sleep, again in contrast to REM sleep, in which they are roughly comparable to the waking state.

Positron emission tomography (PET) imaging studies have also been used to picture the changes that occur in the brain during sleep. In NREM sleep there is a decline in brain activity overall, but also more specifically in some areas of the cerebral cortex and subcortex. This decline progresses as NREM moves into its deeper stages.

BRAIN ACTIVITY DURING SLEEP STAGES

Awake

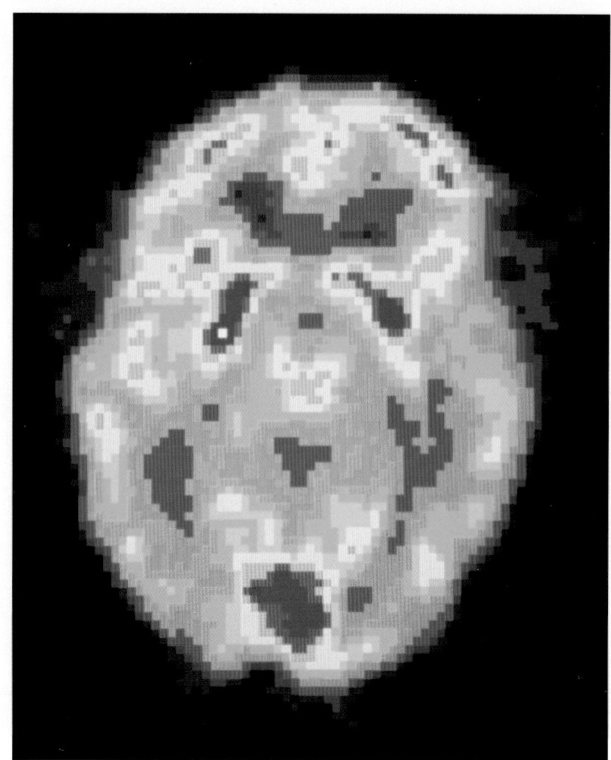

THE AUTONOMIC NERVOUS SYSTEM

In this chapter we primarily deal with the brain and spinal cord, which comprise the central nervous system. There is also an additional peripheral nervous system, and one important part is the autonomic system. The function of the autonomic system is to regulate the body's internal environment. It is made up of two parts, each of which sends fibers from the central nervous system to groups of cells (ganglia) around the body, which in turn send neurons to various organs.

Sympathetic nervous system: This system is traditionally described as one that prepares the body for emergencies and rapid action. Its postganglionic neurons that affect organs secrete epinephrine (adrenaline) or norepinephrine (noradrenaline). For our purposes, it is the portion of the autonomic system which is more associated with REM sleep and arousal.

Parasympathetic nervous system: This is more associated with quieting the body down, sometimes referred to as "rest and digest." Its postganglionic neurons secrete acetylcholine. Its effects are generally opposite to those of the sympathetic system. As noted in the text, in NREM sleep, there is a relative preponderance of parasympathetic activity.

BRAIN CHANGES DURING SLEEP
PET scan images of the brain when awake, in NREM sleep, and in REM. In this procedure, glucose molecules are radioactively "tagged," then injected intravenously. Portions of the brain which are more metabolically active take up more of the tagged glucose, and the amount and locations of photons given off by it are recorded and shown graphically as if seeing slices of the brain. The areas with the most activity are pictured as red, and those with the least are seen as blue. In the waking brain, it can be seen that a number of areas are very active, while in NREM sleep there is a general decline. During REM sleep, the brain becomes metabolically active, in a pattern generally resembling that of wakefulness.

NREM

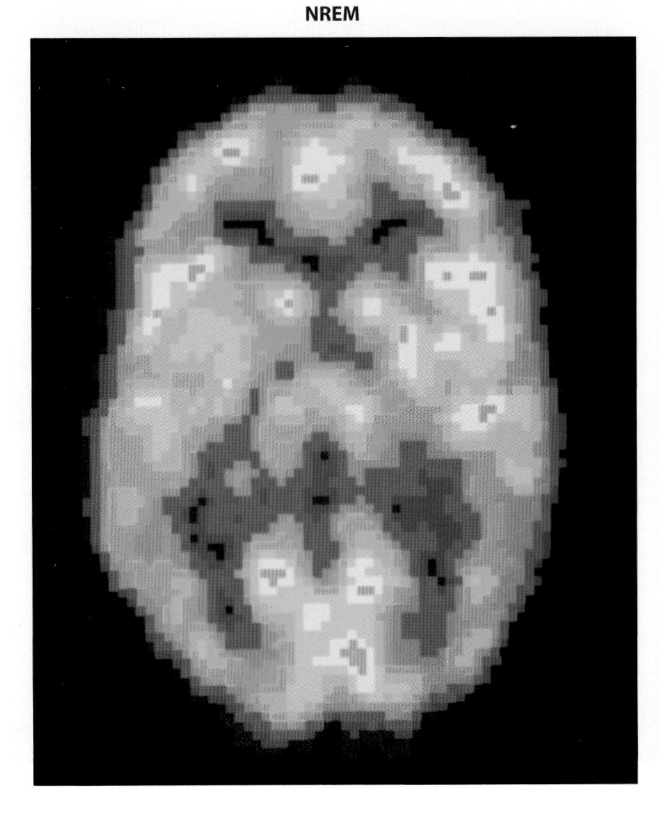

REM

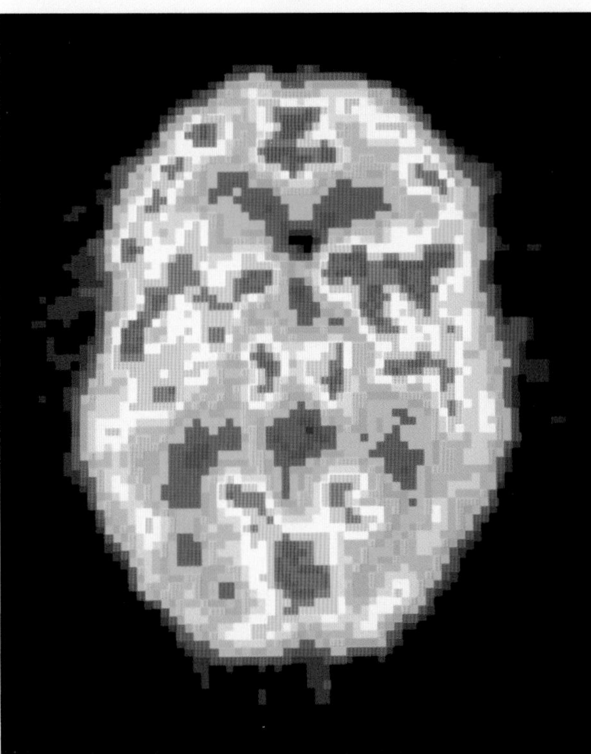

WHAT CHANGES OCCUR IN THE BODY IN REM SLEEP?

The physiology of REM sleep

As mentioned earlier, REM sleep is characterized by low-voltage, mixed-frequency EEG activity, accompanied by the rapid, sharply peaked, conjugate eye movements (as if visually tracking something) seen on the EOG, and loss of muscle tone in the submental EMG (as recognized by the reduced amplitude [height] of the EMG line). The EEG may show "sawtooth waves," a variant of theta rhythm in which there are small notches on the waveform. There are no spindles or K-complexes. Not every epoch of REM sleep has eye movements, but periods of low-voltage, mixed-frequency EEG and low EMG are scored as REM until sleep spindles or K-complexes make their reappearance. When eye movements are present, their total amount (sometimes measured as the percentage of the epoch occupied by eye movements) is referred to as "REM density," which may increase in some disorders, including depression.

The decrease in the submental EMG activity during REM sleep reflects very significant loss of tone of most major muscle groups in the body (one exception is the diaphragm, which is only slightly affected). This is due to the action of an area in the pons (part of the brain stem) known as the peri-locus coeruleus which causes a descending pathway to the spinal cord (the reticulospinal tract) to inhibit neurons in the spinal cord that control muscle tone. As we will see in Chapter Two, lesions of the peri-locus coeruleus area can result in a state in which during REM sleep animals retain their muscle tone, and allow them to act out behaviors that have been interpreted as what they are doing in dreaming.

REM SLEEP INDUCTION

One interesting aspect of REM sleep is that it can be induced in humans by administering certain medications. As we will see in the next section, an important neurochemical in the brain and in the parasympathetic nervous system is acetylcholine. Drugs that enhance acetylcholine function are used for a variety of medical purposes. It had been known since the 1950s that some patients treated with these medicines reported increased dreaming, and in the 1960s it was reported that industrial workers exposed to certain kinds of insecticides with anticholinesterase properties had greatly increased REM sleep. Based on these observations, scientists at the US National Institute of Mental Health infused physostigmine, which increases cholinergic function, intravenously into normal adults during sleep. When this medicine was infused 35 minutes after a subject began N2 sleep, REM sleep would appear within a few minutes; when saline (salt water) was infused at that time, REM would only appear roughly 50–55 minutes later, at its normal time about 90 minutes after sleep onset. These studies seem to say that cholinergic activity is important to the generation of REM sleep in the brain. Indeed, in animal studies, injection of the cholinergic drug carbachol into an area inside the pons known as the nucleus reticularis pontis oralis in animals will cause REM sleep to appear. This general area on the back of the pons contains neurons which are known as "REM-on" cells, some of which release acetylcholine as REM begins.

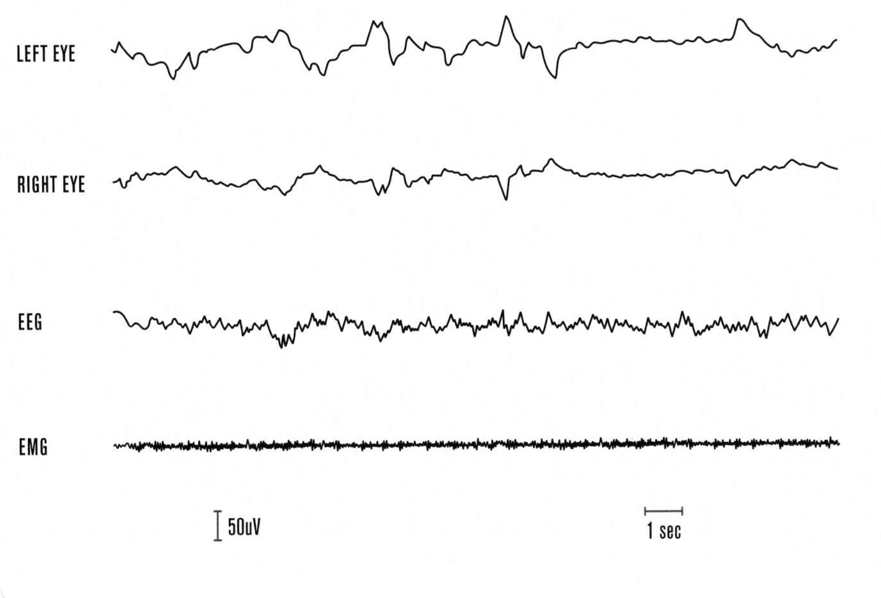

CHANGES DURING REM SLEEP
During REM sleep, the EEG shows low voltage, mixed frequency activity, in many ways similar to N1 sleep. During REM sleep rapid conjugate eye movements can appear. Their conjugate nature, as if the eyes are tracking something, is shown by the way in which they movements of the two eyes are roughly mirror images of each other. The EMG shows minimal muscle tone.

Positron emission tomography (PET) imaging studies of REM sleep have also been performed. Compared to waking, there is relatively less activity in some areas (lateral prefrontal cortex), but greater activation of others including the limbic and paralimbic areas (these parts of the brain are associated with emotions, drives, memory processing, motivation, consciousness, and other processes). Periods of eye movements during REM are particularly associated with activation of areas important in arousal, attention, and emotion.

Although REM sleep is named for the appearance of rapid eye movements, there are actually a wide variety of physiological changes that separate it from both waking and NREM sleep. Many of these are due to alterations in autonomic nervous system control. Respiration becomes irregular, and there is further loss of tone in muscles in the chest wall and throat involved in breathing. The ability to regulate blood carbon dioxide by breathing more air when CO_2 levels rise, which is already slightly reduced in NREM sleep, diminishes much further, and can become almost negligible in REM sleep. One consequence of these multiple changes in ventilatory control is found in the clinical observation that in patients with obstructive sleep apnea (see Chapter Seven, pages 110–11) the most severe apneas tend to occur in REM sleep. A variety of other physiologic changes occur as well. Brain temperature increases, while the typical body responses to outside temperature, such as sweating or shivering, are lost. In effect, we become like cold-blooded animals ("poikilothermic") during REM sleep.

One consequence of altered autonomic activity is that men tend to develop penile tumescence (erections). At one time, measures of tumescence during sleep laboratory studies were used to help separate medical causes of impotence (due, for instance, to vascular disease) from those due to psychological factors such as anxiety. The reasoning was that if a man had tumescence during REM sleep but not when awake, the difficulties were more likely related to psychological reasons. This method has become less widely used in sleep laboratories, partially because the issue turned out to be more complex (for instance, there may be a decline in nocturnal tumescence in depression) and also with the advent of portable home testing devices by urologists.

HOW MUCH SLEEP DO WE GET, AND HOW MUCH DO WE NEED?

Sleep duration and health

The amount of sleep we get varies across the lifetime, from very high amounts of up to 16 hours in newborns, declining across childhood, and leveling off for most of adulthood until old age (see Chapter Five, pages 86–97). There are also individual differences, for which there is a significant genetic component. At least two studies of twins, in Finland and Australia, have indicated that genetic factors account for one third or more of variance in the quality and quantity of sleep. There are also "natural short sleepers" and "natural long sleepers" whose sleep need varies substantially from the average. The former live comfortably with perhaps 4–5 hours of sleep per night, and feel fully awake in the daytime. In less extreme cases, well-known figures such as Napoleon, Thomas Edison, and Margaret Thatcher are said to have gotten by with relatively short sleep, while others including Einstein were reported to sleep longer than average. There have been personality studies of long and short sleepers. One well-known study indicated that adults who sleep less than 6 hours tend to be more hard-working, conventional in their standards, and less artistic than long sleepers (who slept more than 9 hours per night). There was also evidence of what were called variable sleepers, who slept substantially longer during times of stress and worry compared to when life felt more stable.

Adults in the United States were reported to obtain about 8 hours sleep in the 1960s, while more recent Gallup polls showed about 7 hours in 2005. A large Finnish study showed a more modest decline to about 7.3 hours. About one in five adults felt that there was a difference of more than 1 hour between the amount of sleep they needed and the amount they actually had.

Sleep and health

Decreased sleep has been associated with higher mortality, as well as a variety of more specific health problems. There is a higher risk (up to about 12 percent) of mortality in those sleeping less or substantially more than 7 hours. The US Institute of Medicine and Centers for Disease Control and Prevention have noted the association of inadequate sleep with chronic conditions including hypertension, diabetes, depression, and

LONG AND SHORT SLEEPERS
Albert Einstein habitually slept 10 hours a night. In contrast, Napoleon is said to have been a short sleeper, but was also famous for taking short refreshing naps. In addition to his natural tendency toward short sleep, the pressures of war may have limited his sleep time as well. Some have speculated that his questionable judgments at Waterloo were in part related to sleep deprivation.

WAKE-UP TIMES AND BEDTIMES ACROSS 20 COUNTRIES

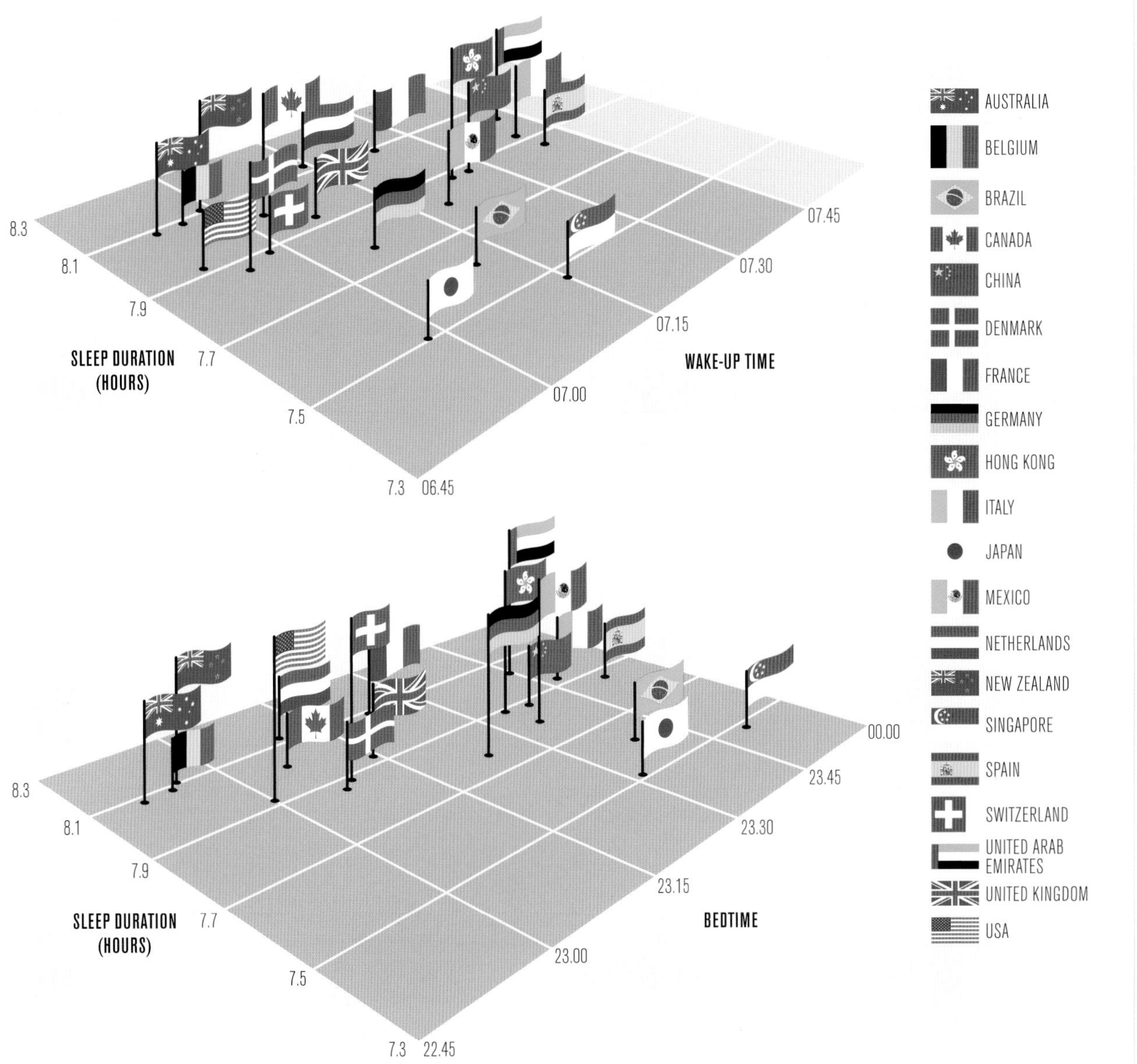

SLEEP DURATION (HOURS)
8.3
8.1
7.9
7.7
7.5
7.3
WAKE-UP TIME
06.45
07.00
07.15
07.30
07.45

SLEEP DURATION (HOURS)
8.3
8.1
7.9
7.7
7.5
7.3
BEDTIME
22.45
23.00
23.15
23.30
23.45
00.00

AUSTRALIA
BELGIUM
BRAZIL
CANADA
CHINA
DENMARK
FRANCE
GERMANY
HONG KONG
ITALY
JAPAN
MEXICO
NETHERLANDS
NEW ZEALAND
SINGAPORE
SPAIN
SWITZERLAND
UNITED ARAB EMIRATES
UNITED KINGDOM
USA

VARIATION IN SLEEPING HABITS AROUND THE WORLD
In this study a smartphone app called ENTRAIN was used to measure bedtime and morning waking times around the world. The wide variation in habits can be seen. Differences in the timing of sleep between countries was related more to differences in bedtime than waking up time. The average bedtime in a given country was more predictive of total amount of sleep than was waking up time. Age had an important effect, insofar as the sleep habits of people over 55 were less variable. It was also found that people who report being exposed to more outdoor light tended to go to bed earlier and have longer sleep durations. Adapted from Walch et al., 2016.

obesity. Some specific examples: short sleepers, both children and adults, are at higher risk for developing obesity, and in adults both shorter and longer sleepers have a higher rate of developing type 2 diabetes. There have been similar studies of hypertension with less clear results. In women, short sleep has been related to an increased likelihood of coronary heart disease. The topic of decreased sleep time and health is dealt with in more detail in Chapter Three. It is tempting to speculate mechanisms for these kinds of findings—for instance, that in those with shorter sleep duration, sleep does not have adequate time to perform its possible health-promoting functions—but it is important to remember that these are associations, not proved causal links. It is possible, for instance, that some unknown third factor causes both shorter sleep and a less desirable health outcome.

It is also important to remember that healthy sleep is a broader concept that involves much more than just not being sleep deprived. The American sleep researcher Daniel Buysse (1958–) has emphasized that it involves duration, continuity of sleep, timing, the ability to maintain good wakefulness, and a subjective sense of satisfaction with sleep. He has defined sleep health as follows: "Sleep health is a multidimensional pattern of sleep/wakefulness,

SLEEP VARIATION ACROSS AGE AND GENDER
Average sleep times from the Bureau of Labor Statistics, American Time Use Survey 2015. Age and gender have important influences on the total amount a person sleeps. There is wide variation between individuals, but this figure gives a general sense of the average in large populations.

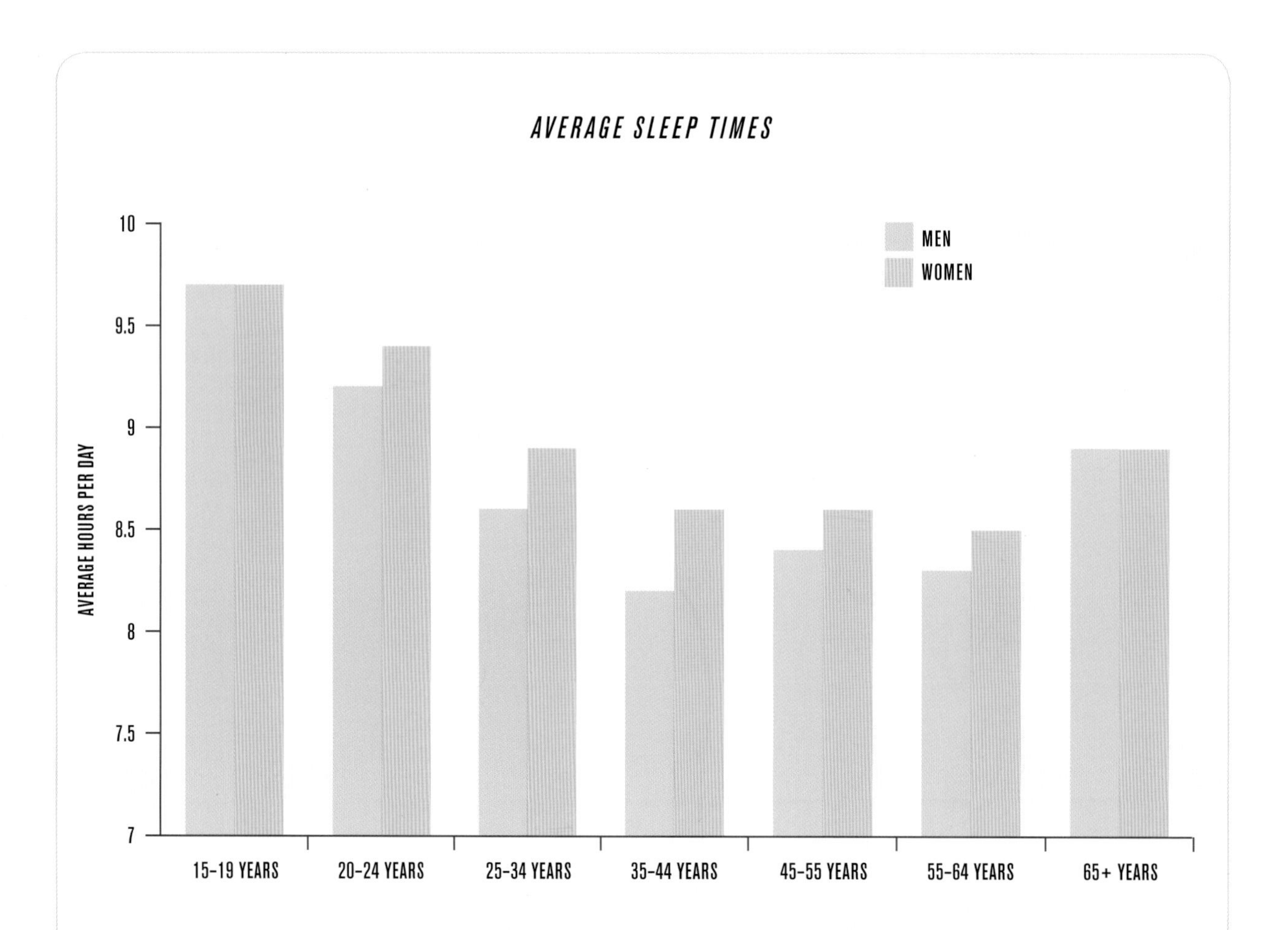

adapted to individual, social and environmental demands, that promotes physical and mental well-being. Good sleep health is characterized by subjective satisfaction, appropriate timing, adequate duration, high efficiency, and sustained alertness during waking hours."

How much sleep do we need?

In the preceding paragraphs we talked about how much sleep we get. A related, but slightly different question, is how much sleep we need. It's difficult to define how much we need, since the functions of sleep are only partially understood. Probably the most common way to look at it is that one major function of sleep is to promote good daytime wakefulness (see Chapter Three, page 70, for discussion of possible functions of sleep); therefore the amount of sleep we need is the amount that results in adequate daytime wakefulness as measured both objectively and by a subjective sense of wakefulness and energy. One way this has been studied is to determine habitual sleep in a person who has no pre-existing deficit. Studies of adults who have been given the opportunity to have several nights of long sleep (and hence to catch up on any sleep they lacked) tend to show that typical sleep is in the area of 7.5–8.5 hours, primarily about 8.17 hours. Sleep deprivation is considered to be a lesser amount of sleep that results in deficits in wakefulness and daytime functioning. People who are sleep deprived develop a "sleep debt," and when given the opportunity, sleep for longer periods as if to make up for the debt. This is an example of a basic regulatory principle of sleep, the "homeostatic drive," and also represents one of the fundamental qualities of sleep, which is that it is self-regulating.

The sleep debt

It is important to understand, though, that having a sleep debt and then compensating through longer sleep is not as simple as owing some money and then paying it back. In the case of money, if you stumble upon some unexpected cash you can take it to the lender, and everyone is happy. In the case of repaying a sleep debt, it is more complicated. For one thing, after acute sleep deprivation, the amount of recovery sleep needed may be less than the total amount that had previously been lost. On the other hand, some complicating factors may limit one's ability to pay off the debt. To give an example, many sleep researchers have had the experience of stumbling home to bed at 7:00 AM after a full night awake studying sleeping patients, going to bed in a sound sleep, then awakening very tired about 11:00 AM or noon. In this case, other factors (primarily the body clock, or circadian mechanisms) prevented sleeping for a longer time, though it was clearly needed. Similarly, if a person is acutely sleep deprived, then allowed to sleep as long as (s)he wants, the first night's sleep will be very long indeed, but measures of wakefulness the next day will show that (s)he is still sleepy to some degree. It will take several days of long sleep until normal wakefulness returns. Doctors in training in anesthesiology, for instance, have been found to be very sleepy, even when not having been on call for two days. We will talk at length about sleep deprivation in Chapter Three, but the important message here is that sleep is self-regulating, through what is known as the homeostatic mechanism.

What are some of the consequences at a physiological level of developing a sleep debt? One possibility is that the brain may have inadequate time to clean itself out. It turns out that the brain has a series of "pipes" which run near blood vessels, into which waste products of metabolism are collected and ultimately flushed from the body. This system is served by glial (non-neuron) cells called astrocytes, and has been dubbed the "glymphatic system." This process is much more efficient during sleep, and it has been speculated that with inadequate sleep the ability of the brain to clean itself is compromised.

WET AND DRY PHYSIOLOGY

Chemical transmitters (Wet) and electrical phenomena (Dry) involved in sleep and waking

In 1918, one of the greatest natural disasters in human history occurred, a worldwide epidemic of a particularly virulent kind of flu. Ultimately, about 500 million people around the world were infected, and about 50–100 million died. It killed more people than the Black Death in the Middle Ages. Coincident with this was a worldwide outbreak of a related illness, encephalitis lethargica, a viral inflammation of the brain which itself caused 500,000 deaths, and often led to severe lasting neurologic injuries.

Sleep—a passive state?

Around the time of the outbreak, Constantin von Economo (1876–1931), an Austrian psychiatrist and neurologist who had studied under Alois Alzheimer and who had served as a pilot in South Tyrol (northern Italy), returned to Vienna to take care of soldiers with neurological disorders. He became interested in some of the encephalitis patients who were left with syndromes of severely excessive sleepiness and others with chronic sleeplessness. He related the sleepiness syndrome to lesions in the brain stem and posterior hypothalamus, and the sleeplessness syndrome to damage to the basal forebrain and striate structures. This was among the first observations that specific brain areas could be associated with sleep and its regulation. In the 1930s, Frederic Bremer (1892–1982), a Belgian neurophysiologist, who had among other things served as a physician in the cavalry in World War I, investigated sleep and waking in cats. He found that cats in whom the cerebrum was unable to receive inputs from the brain stem (known as "cerveau isolé") stayed in a sleep-like state, with the slow, high amplitude EEG waves of sleep, and could not be easily aroused. In contrast, cats in whom the brain stem could not receive input from the spinal cord, but which retained their connections with the cerebrum ("encephale isolé") showed the behavioral and EEG signs of wakefulness. His interpretation was that stimuli from the external world, such as might be brought to the brain by the cranial nerves leading to the brain stem, were necessary for wakefulness. Sleep, then, was perceived to be a kind of resting state, to which an animal reverted in the absence of stimulation. This view was strengthened in the 1940s by the work of the Italian researcher Giuseppe Moruzzi (1888–1973) and the American Horace Winchell Magoun (1907–91), who discovered a diffuse network of cells, running upward from the brain stem to the thalamus, a structure in the midline of the brain lying between the cerebral cortex and the upper part of the brain stem (midbrain). These neurons were activated by sensory input, but appeared not to bring specific sensory information. Rather, a general message of stimulation was being delivered to the cerebral cortex via the thalamus, and via a second more anterior pathway to the hypothalamus and basal forebrain. Electrical stimulation of this "reticular activating system" (RAS) would awaken a sleeping animal. It was thought that, in the absence of such stimulation to the cerebrum, systems in the thalamus and lower brain stem synchronized the firing of neurons in the cerebral cortex, resulting in the slow waves of sleep. Sleep, then, was seen as a kind of passive state occurring in the absence of stimulation.

Passive vs active regulation of sleep

In subsequent years, two types of studies called this passive view of sleep into question. The first was the finding that electrical stimulation of several areas of the brain could induce sleep, instead of wakefulness. The observation that stimulating these areas, which

RETICULAR ACTIVATING SYSTEM (RAS) PATHWAYS

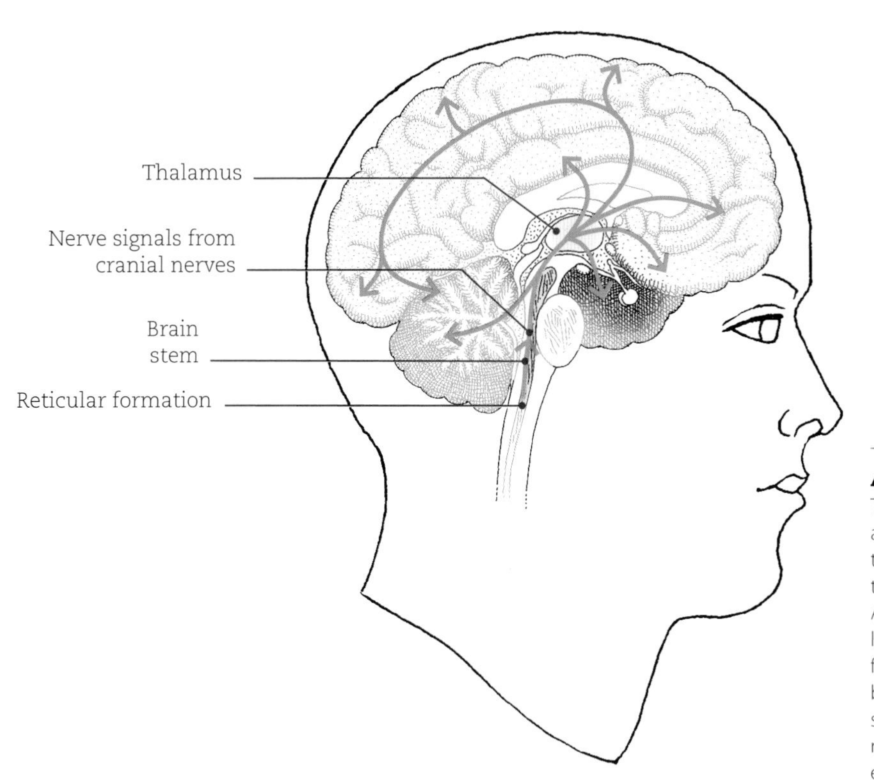

ACTIVATING SIGNALS
The main pathway of the RAS, ascending from the brain stem to the thalamus, from which radiations go to much of the cortex, is seen here. A second, more anterior pathway leads to the hypothalamus and basal forebrain. The RAS, which is stimulated by sensory input, does not provide specific sensory information, but rather a general activating signal; electrical stimulation of portions of the RAS lead to wakefulness.

include parts of the brain stem, thalamus, and basal forebrain, results in sleep indicated that sleep is an actively controlled process, not just a default state that occurs when there is not enough stimulation.

The second blow to the notion of sleep as a passive state came with the discovery of REM sleep. Before that time, Nathaniel Kleitman (in whose laboratory REM was later discovered—see pages 14–15) had visualized sleep as analogous to ice while waking was like water. The subsequent realization that there are two very different sleep states (REM and NREM) whose rhythmic appearance and disappearance is governed by an elaborate regulatory system made it necessary to reconsider the passive model. This, plus the recognition that sleep could be induced by electrical stimulation of certain brain structures, led to the realization that sleep is an actively regulated phenomenon.

When sleep had earlier been thought of as a passive state, most of the theories of how it worked could be thought of in what was sometimes referred to as "dry neurophysiology," that is, in terms of electrical phenomena. For instance, it was possible to speculate that issues of neuronal fatigue in the RAS as well as decreased sensory stimulation predispose to sleep, or that sleep is in some sense a process analogous to recharging batteries. Many of the observations of the active regulation of sleep made this difficult, however. For instance, the duration of the REM–NREM cycle (about 90 minutes), or the REM rebound phenomenon of increased REM following deprivation (which could go on for several nights) were hard to reconcile with the changes in electrical potentials of neurons, which are generally measured in thousandths of a second. It was easier to consider these long-lasting processes in terms of "wet neurophysiology," the study of chemical transmitters in the brain.

NEUROTRANSMITTER PATHWAYS

Neurotransmitters and their role in sleep regulation

In the 1960s, two types of studies demonstrated the benefits of studying neurotransmitters. The first study showed that infusions of neurotransmitter chemicals such as serotonin and norepinephrine into the brain could profoundly affect sleep and waking. The second discovery made it possible to trace the pathways of neurons containing these neurotransmitters in the brain. It was discovered that by exposing some catecholamines (a class of neurotransmitters) to formaldehyde, they would fluoresce, and hence the pathways containing these compounds could be visualized by special microscopes in tissue slices. Prior to this second discovery, it had been very difficult to track pathways, for instance in the very diffuse structures of the RAS.

BRAIN ANATOMY

Important structures for sleep include the brain stem (comprising the medulla, pons, and midbrain), the thalamus (located between the cerebrum and midbrain), and hypothalamus (an almond-sized structure located just above the brain stem). The hypothalamus contains a variety of nuclei regulating sleep and wakefulness, and also serves as an integrating center, relating sleep to other physiological processes including temperature regulation, and endocrine and circadian mechanisms.

Serotonin

It was found that serotonin-containing neurons were largely concentrated in midline structures known as the dorsal raphe nuclei in the lower midbrain and pons. From there, nerve fibers travel upward to widespread areas of the diencephalon (part of the forebrain including the hypothalamus and thalamus) and the cortex. Serotonin has a variety of functions in the nervous system, affecting not only sleep–wakefulness, but also mood, thermoregulation, appetite, memory, and activity in the suprachiasmatic nucleus (the main body clock). Peripherally it is involved in regulating motility of the intestines.

SLEEP CONTROLS IN THE BRAIN

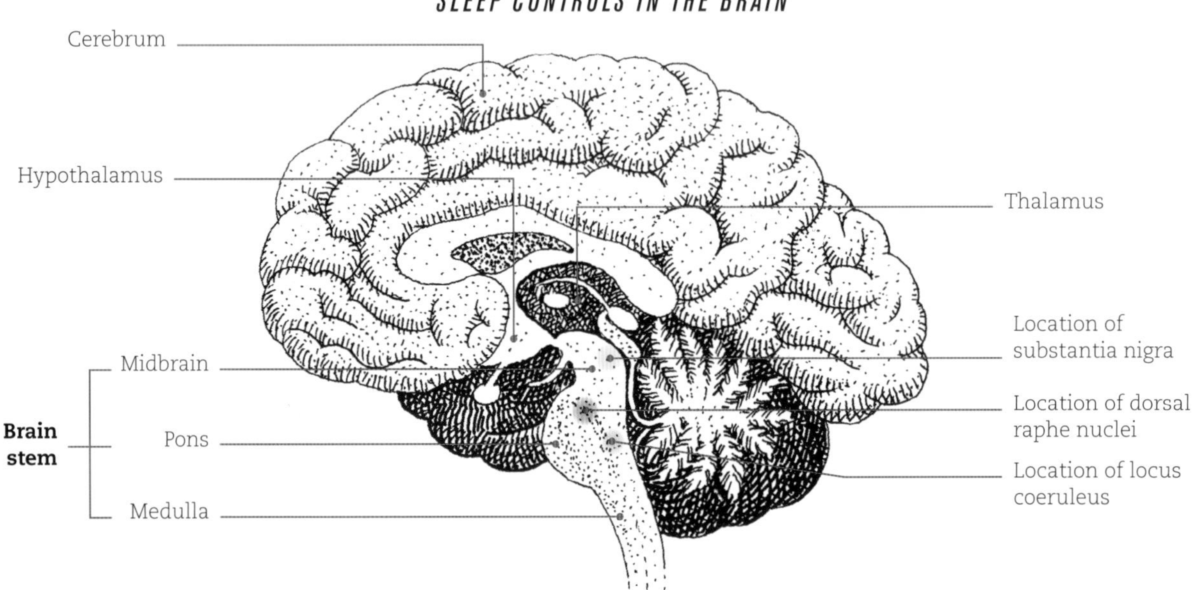

SEROTONIN PATHWAYS

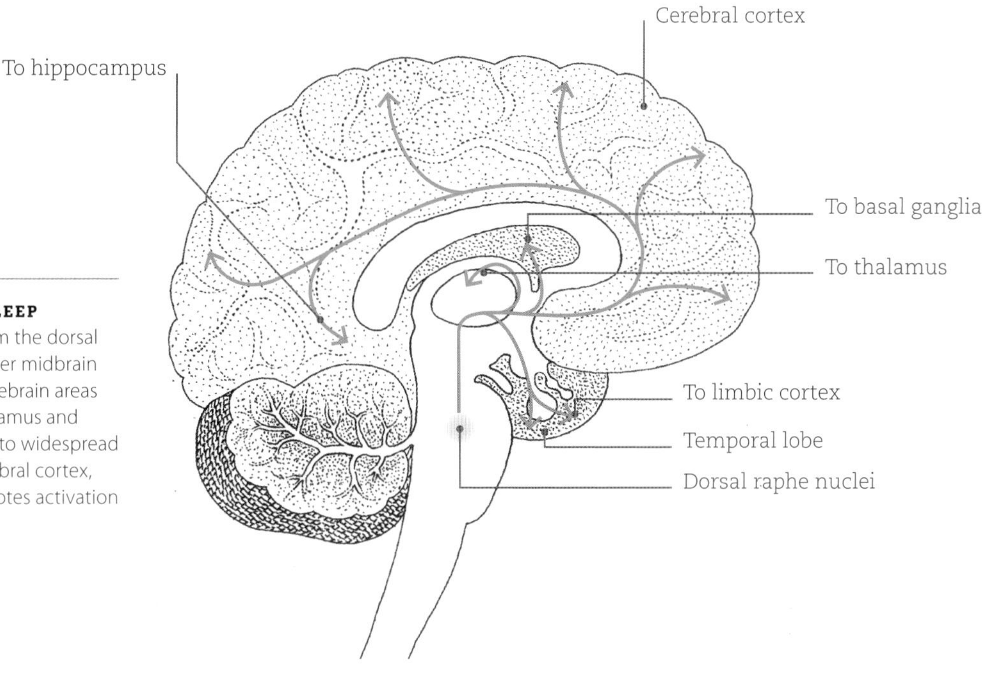

SEROTONIN AND SLEEP
Serotonergic fibers from the dorsal raphe nuclei in the lower midbrain and pons ascend to forebrain areas including the hypothalamus and thalamus, and then on to widespread distribution in the cerebral cortex, where serotonin promotes activation and wakefulness.

Norepinephrine

Norepinephrine (noradrenaline) neurons are widely distributed in the reticular formation, with heaviest concentrations in the locus coeruleus (the "blue nucleus") in the pons. Ascending fibers go to widespread areas of the diencephalon and forebrain, as well as to the cerebellum. In the central nervous system it is thought to play a role in sleep/wakefulness as well as in the state of vigilant concentration. Along with epinephrine (adrenaline), it helps regulate the fight-or-flight response, increasing heart rate and blood flow to muscles.

Dopamine

Dopamine-containing neurons are concentrated in midbrain nuclei known as the substantia nigra ("black substance"), adjacent areas of the midbrain, and the basal and lateral hypothalamus. They have highest activity in waking and REM sleep, and dopamine is released in the cortex more in waking than sleep. Dopamine is thought to have an important role in cognitive alertness.

Glutamate

The amino acid glutamate is an excitatory neurotransmitter found throughout the brain and particularly in the pontine and midbrain reticular formation. Glutamate pathways are not as well worked out. Neurons employing glutamate descend from the brain stem to the spinal cord, leading to the muscle relaxation seen in REM sleep. Application of glutamate to a number of brain sites seems to lead toward wakefulness.

Acetylcholine

Studies of acetylcholine did not have the advantage of histofluorescence, but neurons rich in acetylcholine were found in the dorsolateral pontine and midbrain reticular formation as well as the basal forebrain. It plays an important role in activation of the cortex and behavioral arousal. Release of acetylcholine from the cortex is greatest in waking and REM sleep, and lower in NREM sleep.

NEUROTRANSMITTER EFFECTS ON SLEEP

How they function in regulating sleep and wakefulness

For most of the neurotransmitters we've mentioned so far, ascending pathways rise from the brain stem to influence sleep and waking. Ultimately it was found that waking is promoted by a variety of neurotransmitters, including norepinephrine, acetylcholine, histamine, glutamate, and orexin/hypocretins. The role of serotonin is more complex. Although stimulation of some subtypes (the "5-HT_{2A} receptor") may promote NREM sleep, the overall effect of drugs that increase serotonergic activity is to promote wakefulness. Many of these neurotransmitters that promote wakefulness also have effects on REM sleep (see table below).

Neurotransmitters that promote NREM sleep include GABA (gamma-aminobutyric acid), the most common inhibitory neurotransmitter (reducing excitability) in the nervous system, and adenosine. As previously mentioned, acetylcholine appears to be involved in the promotion of both wakefulness and REM sleep. Orexin/hypocretins are peptides (molecules consisting of chains of amino acids) from neurons in the lateral hypothalamus, with projections to the brain stem, other parts of the hypothalamus, the limbic system, thalamus, and more widespread parts of the brain. They potently tilt the nervous system toward wakefulness, and defects in orexin function are thought to be related to the genesis of the illness narcolepsy (see Chapter Seven, pages 124–5). Increased orexin release during sleep deprivation is also thought to be related to increased appetite during sleep deprivation (see Chapter Three, page 64).

Histamine-containing neurons are located in part of the hypothalamus known as the tuberomamillary nucleus in the posterior hypothalamus, and send projections into various areas of the hypothalamus, cortex, and brain stem. Histaminergic activity promotes wakefulness; antagonists of the histamine type 1 receptor that cross the blood–brain barrier (such as diphenhydramine) cause drowsiness, and are often sold as over-the-counter sleep aids (see Chapter Eight, pages 168–9).

Models of sleep–wake regulation

Neurons containing the neurochemicals described in the last section interact in a complex way to cause the transition from waking to NREM sleep, and to produce the alternating cycles of REM and NREM. The switching mechanism between waking and NREM is thought be controlled by the balance of activity between neurons in two nuclei of the hypothalamus (ventrolateral preoptic, or VLPO, and median preoptic) which release GABA, and wake-promoting neurons which release serotonin, norepinephrine, histamine, and orexin/hypocretins. Neurons in the VLPO and median preoptic are more

NEUROTRANSMITTERS/NEUROMODULATORS PROMOTING WAKEFULNESS AND SLEEP

Those which promote wakefulness and REM:	Those which promote wakefulness while inhibiting REM:	Those which promote NREM sleep:
Acetylcholine	Serotonin (see text)	GABA
Dopamine	Norepinephrine	Adenosine
Glutamate	Histamine	
	Orexin/hypocretins	

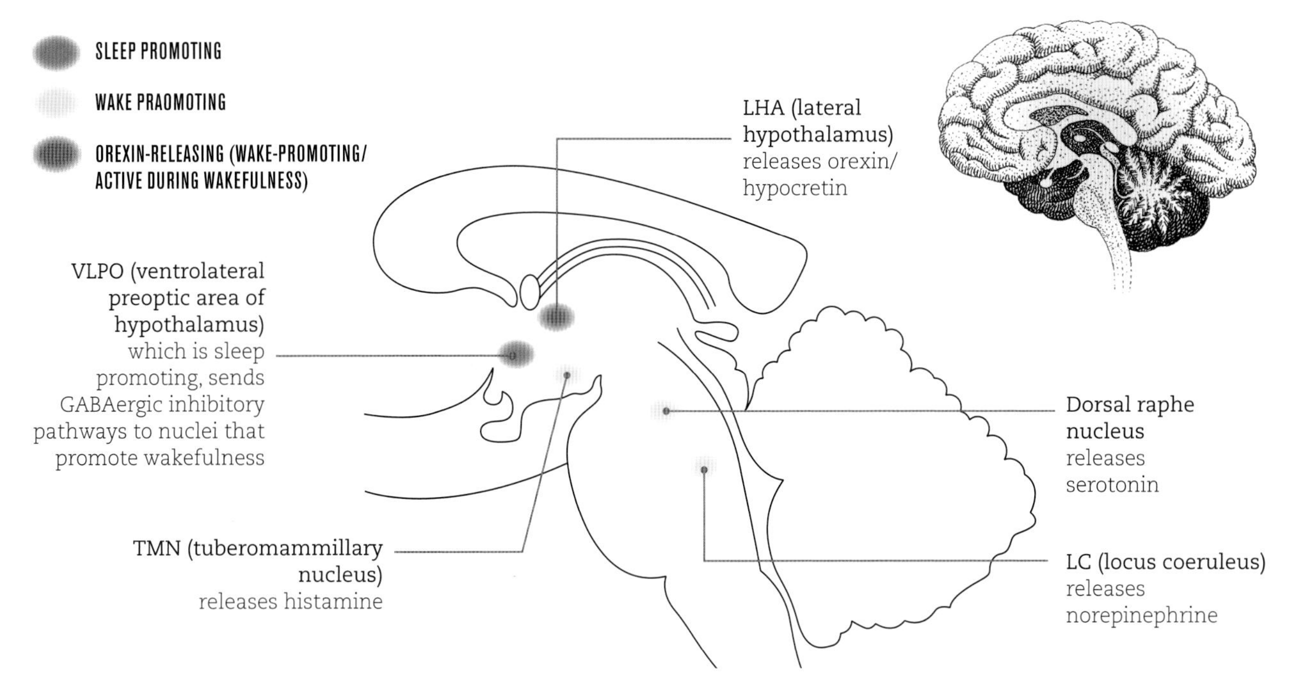

NEUROTRANSMITTERS IN THE SLEEP–WAKE CYCLE
The alternating periods of wakefulness and sleep arise from the interaction of nuclei whose neurotransmitters promote the two states. Neurotransmitters including serotonin from the dorsal raphe, norepinephrine from the locus coeruleus and histamine from the TMN, are involved in pathways with broad distribution to the cerebral cortex where they promote activation, and additionally some of these arousal nuclei send inhibitory signals to the sleep-promoting VLPO. The VLPO and related hypothalamic nuclei, in turn send GABAergic inhibitory signals to the various arousal nuclei. The result is a cyclic process of waking and sleep. There are a number of other important neurotransmitters/ neuromodulators involved, including orexin/hypocretin pathways from the LHA which are active during waking and participate in stimulating the arousal nuclei. In addition to promoting wakefulness, this may help prevent inappropriate transitions into sleep (see pages 124–5). Adenosine, which is produced as a result of energy metabolism in neurons and glial cells, accumulates during wakefulness, promoting sleep, and concentrations decline during sleep. Other processes which contribute to regulating sleep and wakefulness include the circadian system.

active in sleep, while those in the wake-promoting area are more active during waking. These two systems mutually inhibit each other. In later sections we will discuss some of the regulatory mechanisms which help tip the system in one way or the other, including homeostatic factors and time of day (circadian influence), as well as the effects of sleeping pills on this process.

NEUROTRANSMITTER TERMINOLOGY

Neurons are sometimes described by the neurotransmitter they use to communicate with other neurons. The terms for these, and the neurotransmitters they refer to, are:

Cholinergic: uses acetylcholine. Example: laterodorsal tegmental nucleus (LDT) and pedunculopontine tegmental nucleus (PPT) which are the cholinergic REM-on centers in the reciprocal interaction model.

Serotonergic: Serotonin. An example would be the dorsal raphe nuclei in the brain stem, which are part of the REM-off neurons in the reciprocal interaction model.

Noradrenergic: Norepinephrine. Example: the locus coeruleus in the brain stem, which represents part of REM-off neurons in the reciprocal interaction model.

Aminergic: a broader term for the biogenic amines, including serotonin, norepinephrine, and dopamine.

Glutaminergic: Glutamine. Example: Neurons descending from the REM-on centers in the flip-flop model, which modulate muscle relaxation during REM sleep.

GABAergic: GABA. Examples: The neurons in the REM-on and REM-off centers in the flip-flop model for REM–NREM regulation.

Regulation of the REM–NREM cycle

The regulation of the cycle of REM and NREM sleep is also thought to be due to a complex interaction of opposing sets of neurons. Various models have presented ways it might work. One, known as the "reciprocal interaction model"and associated with the American neuroscientists Allan Hobson (1933–) and Robert McCarley (1937–), was first developed in the 1970s and has continued to evolve, with many revisions over the years. In the most basic form of this model, "REM-on cells," cholinergic neurons in the mesopontine tegmental area facilitate REM sleep by exciting glutaminergic neurons in the reticular formation. (Note: Tegmental refers to the posterior, or dorsal, surface of the part of the brain stem known as the pons; mesopontine refers to the upper part of this area, close to the junction of the pons and midbrain.)

They interact in a complex manner with neurons containing serotonin and norepinephrine in the dorsal raphe and locus coeruleus nuclei respectively ("REM-off cells") to produce the cyclic alternation of REM and NREM sleep. Other neurotransmitter systems that affect the process include adenosine, GABA, and orexin/hypocretin.

ORIGINAL RECIPROCAL INTERACTION MODEL

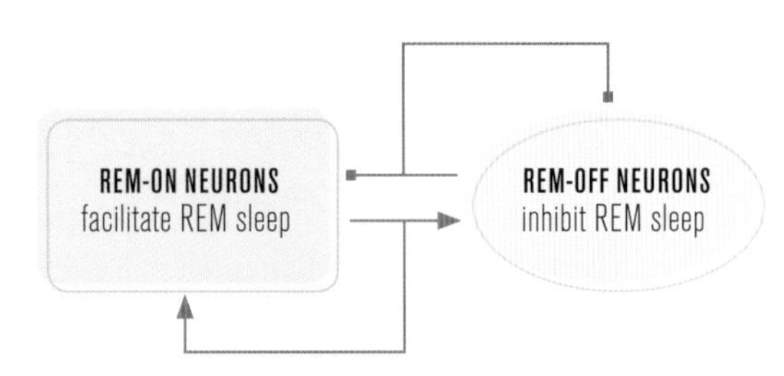

MODIFIED RECIPROCAL INTERACTION MODEL

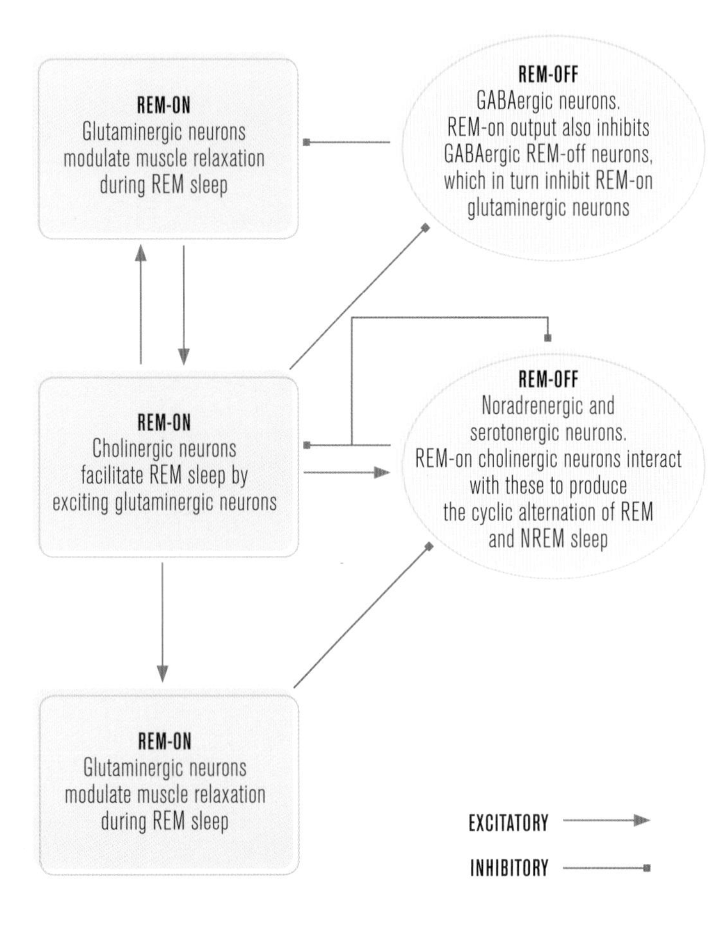

RECIPCROCAL INTERACTION MODEL

The original model developed by Robert McCarley and Alan Hobson (top) shows the complex interaction of cholinergic neurons which promote REM sleep ("REM-on cells") and aminergic neurons which inhibit REM sleep ("REM-off cells"), resulting in the REM–NREM cycle. The model was later expanded (right) to include among other things the role of GABAergic interneurons and the effects of cholinergic REM-on cells on glutaminergic neurons in the pontine reticular formation, which also promote REM sleep and modulate muscle relaxation during REM. As the REM period lengthens, cholinergic REM-on cells also begin to stimulate the REM-off cells, leading to the end of the REM period and beginning of NREM sleep. During NREM sleep, the aminergic REM-off cells inhibit the cholinergic REM-on cells. The REM-off cells are also self-inhibiting, and during NREM their activity eventually declines, releasing the REM-on cells, and leading to the next REM period. Adapted from Brown et al (2012).

In an alternative "flip-flop" model associated with the American neuroscientist Clifford Saper (1952–), the mutual interaction of two sets of GABAergic neurons in the mesopontine tegmentum regulate the transitions between REM and NREM sleep. (The cholinergic and serotonergic centers described in the previous model are considered to be modulators, rather than generators, of REM sleep.) "REM-on cells," as the name indicates, are more active in REM; in addition, glutaminergic neurons in the REM-on area send descending signals to the spinal cord leading to the muscle relaxation of REM sleep. REM-off neurons in areas known as the periaqueductal gray and lateral pontine tegmentum are active during NREM sleep. The REM-on and REM-off neurons interact in a mutually inhibitory way, and their balance of activity is influenced by other neurochemical pathways which include serotonin and acetylcholine. The result of these interactions is the REM–NREM cycle.

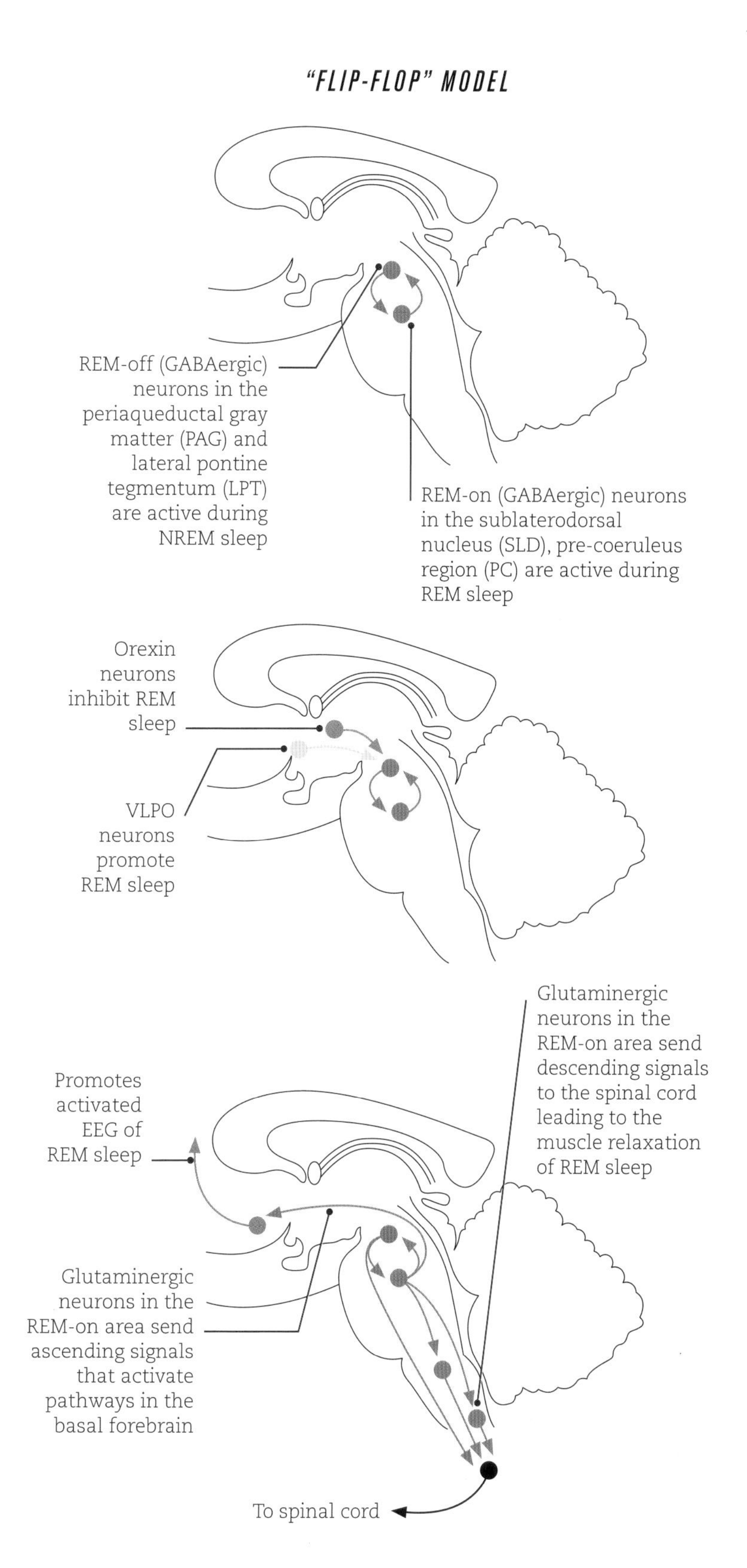

FLIP-FLOP MECHANISM

In contrast to the reciprocal interaction model, the flip-flop model developed by Clifford Saper and colleagues emphasizes the role of non-cholinergic and non-aminergic neurons in the pontine tegmentum in generating the REM–NREM cycle. In this view (as seen in the top diagram) two groups of GABAergic REM-on and REM-off neurons which are mutually inhibitory play a key role in generating the alternating appearance of REM and NREM. (In this model, the cholinergic and aminergic neurons that are the focus of the reciprocal interaction model are considered to be REM modulators rather than generators or inhibitors.) More elaborate versions (middle) show the influence of other systems including pathways from orexin neurons which inhibit, and VLPO neurons which stimulate REM. In the bottom figure, two populations of glutaminergic neurons in the REM-on area are shown to modulate the EEG activation and muscle atonia which are characteristic of REM. The recognition that there are two different sets of neurons and pathways regulating these two processes may help explain how they sometimes become dissociated in conditions such as REM behavior disorder (see page 133). Adapted from Saper et al (2010).

WHAT ARE DREAMS?

Images, feelings, or thoughts that appear during sleep

The richness of the dream experience could fill many a book, but here we are going to focus on the study of dreams from the point of view of neuroscience. Dreams are usually studied in the laboratory by awakening subjects in various stages of sleep and asking them about their mental processes just before the awakening. Several generalizations come from this work. Dream reports are more common late at night. Dreams in REM sleep tend to be more visual ("hallucinatory"), with more emotion, more story-like (have a narrative), and more fictive. In contrast, NREM mentation is more logical (sometimes called "directive thinking") and less visual. In the rest of this chapter, when we say "dreaming," unless specified otherwise we will be referring to REM dreams, which are also to be distinguished from the loosening of thought and occasional images as one first is going off to sleep (sometimes referred to as "reverie").

In the 1950s, when REM was first discovered, the general view about dreams was taken from the work of Sigmund Freud (Austrian, 1856–1939) the founder of psychoanalysis. In his approach, each of us has to wrestle with basic drives that we have to control to prevent getting into trouble. In Freud's view, one function of dreams was to give some expression to these unacceptable drives so that we get at least partial gratification, without allowing them to excessively appear in our daytime lives. Later studies of children and adolescents called this notion into question, suggesting that instead of a process of drive regulation, dreaming seemed to be a cognitive skill that developed in parallel with a child's level of daytime cognitive function. The content of dreams appears to be a mixture of fragments of memories of the preceding day, memories from the more distant past, well-known facts or procedures ("semantic memory"), and stored sensory or motor images.

Of the many ways of understanding the function of dreams, two that appear most likely emphasize the role of dreams in memory processes and in mood regulation. They do not seem to be incompatible, and it is very possible that dreaming, like many physiological processes (breathing, for example), may serve more than one purpose. For example, just as breathing provides the body with oxygen for the blood, it also serves as a mechanism for talking, conveying emotional states ("heavy breathing"), or playing the trumpet. Similarly, dreaming (and sleep itself) may serve more than one purpose.

Memory-related viewpoint

This view, emphasized by the American sleep researcher Robert Stickgold (1946–) and others, suggests that dreaming brings together visual, perceptual, and emotional memories. In contrast, NREM dreaming might be more involved in processes involving the part of the brain known as the hippocampus, absorbing declarative information (such as facts and verbal knowledge) and moving through spatial dimensions. NREM dreams are more influenced by the previous day's experience, while REM dreams are more driven by past memories and knowledge. In this approach, dreaming is the conscious representation of memory and emotional processing that are going on "offline" while we sleep.

Mood-regulatory viewpoint

In this approach, developed with varying emphases by the American sleep researchers Rosalind Cartwright (1922–), Ernest Hartmann (1934–2013), and others, dreams serve to absorb and deal with emotional experiences of the day by relating them to past experiences and integrating them into one's self-image. In the years following the discovery of REM sleep, studies that paved the way for a mood-regulatory viewpoint included observations that

most (95 percent) of dreams involved the dreamer as the main character, and that unpleasant emotions (fear, anxiety, anger) were twice as common as pleasant feelings. Mood-rating scales given before and after sleep in normal volunteers indicated an increase in happiness after a night's sleep. In this view, dreams take the material from moderately emotional events of the day and integrate them with past experiences, in such a way as to help maintain one's self-image. In contrast, if emotional experiences are overwhelming as in severe trauma, dreams become more negative across the course of the night, and the mood-regulating function is impaired. This may also happen in the long term, when REM sleep is disturbed by sleep disorders such as REM behavior disorder, NREM parasomnias (sleepwalking) or post-traumatic stress disorder (PTSD). Dreams may contain such frightening material that they interrupt sleep, with vivid recall of the emotional experience. These nightmares occur in two to six percent of the population, and are more common in children, and in adults with various psychiatric disorders, including PTSD. The mood-regulatory viewpoint is strengthened by imaging studies during sleep, which indicate that during REM there is increased activation of the integrative visual cortex and limbic system, and decreased activation in parts of the prefrontal or parietal cortex. In summary, functions of dreaming are not fully elucidated, but among them may be roles in memory processes and regulation of mood. These processes may be disturbed by some sleep and psychiatric disorders, as we will see in later chapters.

DREAM WORLDS

Henri Rousseau (1844–1910), who never ventured outside France, was fascinated by exotic settings and painted over 25 jungle pictures. In this painting *The Dream* (1910) he combines images of his mistress, a snake charmer, animals, and exotic plants including the lotus, famous in Greek mythology for producing sleep and forgetfulness. The painting captures the visual nature of dreams as well as the frequent combining of seemingly disparate objects and characters.

Chapter Two

SLEEP IN ANIMALS

Sleep is found throughout the animal kingdom. Some scientists have emphasized that it is a kind of adaptive period of quiescence, a way in which to save energy and avoid being exposed to predators when there are not more pressing activities such as eating and reproducing. Since one aspect of sleep is decreased responsiveness, others speculate that sleep additionally plays some more fundamental role(s) as it has persisted in evolution, while having the potential disadvantage of making an animal less aware of its surroundings. In this chapter we will see how sleep-like behaviors evolved from some of the earliest multicellular organisms, insects, amphibians, and fish, and that beginning with birds there is clear evidence of alternating slow-wave and REM sleep. Some birds and mammals have adapted their nervous systems to sleep in challenging environments such as flying or swimming by sleeping with only one half of their brains at a time. It is possible that animals dream, and some experimental situations have even given glimpses into what they might dream about.

IS SLEEP FOUND IN ALL ANIMALS?

A universal behavior

Sleep in animals has long been a source of fascination. This perhaps stems from the common experience of having a pet share the bedroom or indeed the bed. Many of us have observed dogs vocalizing and moving their limbs during REM sleep. To the scientist, animal sleep also has great interest for at least two reasons. First, some animals have illnesses similar to humans, such as that seen in dogs which manifest symptoms of the disorder of excessive sleepiness known as narcolepsy–cataplexy (see pages 124–7), and thus provide a model for study. Second, a look at how sleep has developed throughout the animal kingdom gives some insights into the importance and functions of sleep.

Sleep seems to have a very crucial role in maintaining life, as it appears very early in evolution, and has grown increasingly more complex in parallel to the development of more elaborate nervous systems. In some animal studies, extremely prolonged sleep deprivation, and alternatively selective deprivation of REM sleep, ultimately lead to death (see pages 54–5). One way to look for insights into the role sleep plays in life is to examine how it developed evolutionarily.

A basic question we might consider is whether all organisms sleep. Aristotle (384–322 BCE), the ancient Greek philosopher who was among other things one of the first great naturalists, concluded in On Sleep and Sleeplessness that, "Almost all other animals are clearly observed to partake in sleep, whether they are aquatic, aerial, or terrestrial." Twenty-three hundred years later the data seem to indicate that he was right on the mark. Although the very simplest single-cell organisms such as bacteria do not show sleep-like behavior, some small multicellular organisms do, and—with a few intriguing and arguable exceptions—once nervous systems developed beyond a basic point of complexity, sleep is recognizable in virtually all creatures. In order to explore this, a recap follows of the basic qualities of sleep, so that we can consider whether they are present in different organisms. As was discussed in Chapter One, some essential aspects of sleep are that it is a period of recurring behavioral quiescence which is reversible (hence distinguishing it from coma), and with significantly reduced responsiveness to the external world (which helps separate it from quiet wakefulness). As was mentioned, it is regulated in part by what is known as a homeostatic mechanism, that is, if you temporarily deprive an organism of sleep, it will later have a compensatory "rebound" increase in quantity or depth, as if to make up for the loss. Finally, it is characterized by decreased consciousness. However, there is no way of assessing consciousness in animals (aside from an intriguing indirect glimpse of inferred dreaming in cats which will be discussed later), but all the other qualities just mentioned can be used to assess sleep-like behavior.

First, we will look at the intriguing exceptions: creatures for which the claim has been made that they do not sleep; among these are the cockroach, zebrafish, and bullfrog.

Cockroaches

It has been observed that cockroaches have periodic episodes of quiescence and decreased responsiveness to stimuli. Although some scientists concluded that there is no evidence for homeostatic regulation, other studies have shown an increase in duration of the rest period following deprivation. Additionally, following deprivation they slip into the quiescent phase more rapidly, and they have an increased degree of immobility. During recovery following prevention of rest periods for 12 hours, they are more likely to assume a body position (horizontal, with antennae horizontal and touching the floor) thought to be associated with a lower level of arousal, though these data are equivocal. The rest periods appear to be vital to cockroaches, as when they are deprived of them for prolonged periods they have increased likelihood of death. This suggests that cockroaches have a kind of "sleep-like" state.

Zebrafish

Zebrafish have periods of becoming motorically quiet at night, with characteristic posture in the adults (a drooping caudal fin). In contrast to the cockroach, they have clear compensatory increases in the duration of subsequent rest periods following deprivation by stimulation such as tapping the tank. As they enter older age (up to four years) their rest–activity rhythm weakens and rest periods become shorter. Adults have some evidence for requiring a stronger acoustic or electrical stimulus to cause arousal behavior after about six seconds of immobility (a "higher arousal threshold"), which is one of the features that distinguishes sleep from resting. These features suggest that the zebrafish exhibit sleep-like behavior.

QUIESCENT COCKROACHES
Most of the pest species of cockroach are primarily active at night during the first few hours after the lights are turned out. Cockroaches also have periodic episodes of quiescence, and following deprivation of this resting state they return to it more rapidly. These quiescent periods appear to be necessary for life, as after prolonged periods of deprivation they develop an increase in metabolic rate and have a higher mortality. It has been speculated that to some degree—and that degree is uncertain—this is analogous to what happens to rats who are sleep-deprived for prolonged periods of time

DO ZEBRAFISH SLEEP?
Zebrafish, like cockroaches, have periodic episodes of quiescence, with compensatory increases in duration after deprivation. In addition, adults in this quiescent state require a greater stimulus for arousal, a quality associated with sleep. The genome of zebrafish has been well characterized and, surprisingly, overlaps substantially with that of humans. They are being studied in relation to genetic aspects of cancer as well as diseases of the heart and nervous system.

Bullfrogs

It has been claimed that captive bullfrogs (*Rana catesbeiana*) do not sleep, and indeed many have minimal sleep-like behavior. Some do have periods of quiescence, though during these periods they can open both eyes, their responsivity to stimuli (measured by respiratory changes in response to small shocks to the skin) is not significantly decreased, and there are no unequivocal EEG changes. It has been speculated that this makes them more available to respond to possible threats in the environment. Whether these quiescent periods are under homeostatic control (that is, whether there are prolonged quiescent periods following deprivation) has not been clearly determined. The jury remains out.

There are other examples out there but, in summary, to some (but not all) researchers, it has not been convincingly demonstrated that some animals completely go without sleep or sleep-like states with qualities including homeostatic regulation and increased arousal threshold. Others question the usual definition of qualities of sleep, suggesting, for instance, that homeostatic control might not be essential to a universal description of sleep in all species. It may be that the rest periods we have described in cockroaches, zebrafish, and bullfrogs are early forms of a sleep-like state that emerges more fully in more advanced organisms. Nor is it clear whether sleep developed at a single point in evolution, or at multiple points in different species in order to serve some important function(s).

RESTING BULLFROG

Bullfrogs appear to have minimal sleep-like behavior. During their resting periods they are able to open their eyes and respond to the environment as easily as when they are mobile. This may be a survival mechanism which is activated to deal with potential threats in their surroundings. In contrast, the tree frog (*Hyla septentrionalis*) needs more stimulus to become alert when in the resting phase, a quality more in common with sleep in mammals.

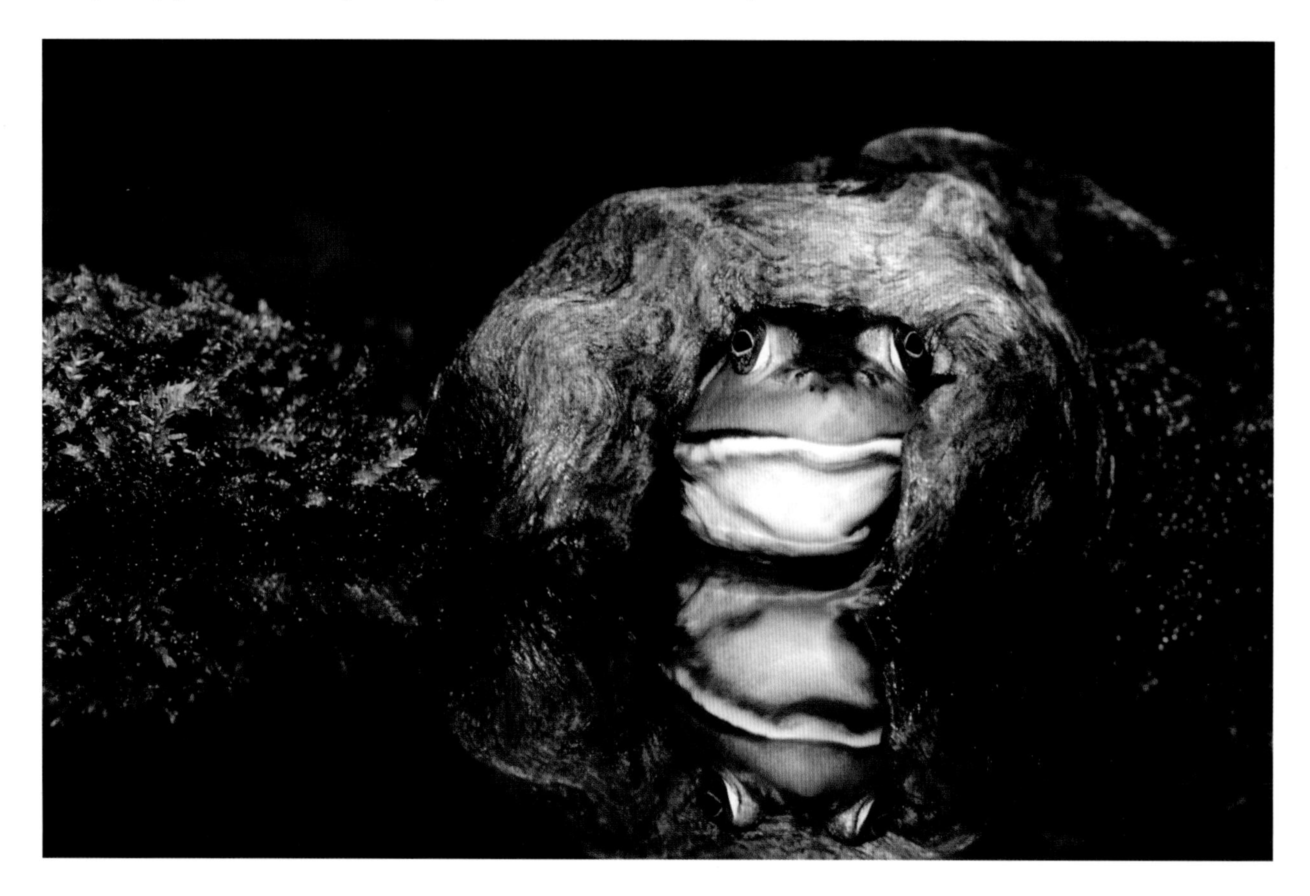

HOW HAS SLEEP DEVELOPED IN THE ANIMAL KINGDOM?

The evolution of sleep

As we mentioned earlier, there is no evidence of sleep-like behaviors in single-celled organisms, but daily periods of inactivity appear in small multicellular organisms known as cyanobacteria and protists. Periods of quiescence and decreased responsiveness to surroundings are evident in worms, bees, and scorpions. In Chapter Four on circadian rhythms (rhythms associated with the body clock) we will discuss sleep in fruit flies, which are of particular interest because they have been found to have genes regulating their body clocks and ability to have recovery sleep after sleep deprivation. We earlier mentioned cockroaches, which when deprived of their daily rest period, subsequently generally display a compensatory depth of immobility and increased rapidity of entering this rest phase. Observations such as this need for recovery, as well as responsiveness to light and dark and time of day, may suggest that periodic episodes of rest represent a state from which sleep as we usually think of it ultimately evolved.

Fish in general have periods of rest thought to represent sleep-like behavior. Catfish have sleep-like episodes characterized by decreased movement and the appearance of possible slow waves on the EEG. In perch and carp, deprivation of sleep by bright lights during the dark period (when they usually rest) results in a subsequent compensatory increase in resting behavior, and in perch this is proportionate to the duration of the prior rest deprivation. We have already mentioned the case of zebrafish (see page 45), who have a rebound increase in sleep-like behavior after deprivation.

Some reptiles including lizards, turtles, and crocodiles are thought to have four states: active and quiet wakefulness, and sleep stages 1 and 2. During sleep, high-amplitude arrhythmic spikes appear on the EEG. After deprivation there is evidence of a recovery sleep phenomenon. It has generally been thought that reptiles do not have REM sleep, but there is now some evidence that a particular lizard, the Australian bearded dragon, may have alternating patterns of slow waves and REM every 80 seconds or so, for up to 6–10 hours. If this finding is confirmed by being replicated in other laboratories, it would suggest that the brain structures necessary for sleep stages may have evolved earlier in evolution than previously had been thought.

AUSTRALIAN BEARDED DRAGON

Reports have been inconsistent on whether REM sleep appears in reptiles, and indeed studies in the same species have differed in results. One reptile for which the case has been made for REM sleep is the Australian Bearded Dragon, which during sleep appears to have alternating patterns of slow waves and REM.

Birds clearly show both slow wave and REM sleep, though in birds the REM periods are much shorter than in mammals, and in general there is less difference between the sleep stages. A case in point is the sleep of the zebra finch, as seen below.

Mallard ducks and pigeons may have brief moments, measured in seconds, in which one side of their brain seems asleep while the other is more cognizant of the environment. Ducks show less of this phenomenon when placed in the center of a group of other ducks, and more of it when at the periphery. This may imply that the one-sided sleep mechanism (known as "unihemispheric sleep") is advantageous in allowing more awareness of the environment in more potentially dangerous locations. It has been speculated that this phenomenon may make it possible to obtain some of the benefits of sleep while also being alert when in the air, a phenomenon that is much more fully developed in some cetaceans (aquatic mammals including whales and dolphins) when under water.

In mammals, there is no clear case of animals which do not sleep, though some can postpone sleep for extended times. The amount of sleep decreases as body mass increases. Elephants, for instance, sleep about 4–6 hours per day. In contrast, total sleep in mammals is not very much related to brain mass. Some scientists have asserted that in mammals and birds, the best predictor of total sleep time is related to their eating habits. Carnivores tend to sleep more than omnivores, who in turn sleep more than herbivores. Mammalian predators tend to sleep longer than prey; having less sleep would presumably be a survival advantage in the prey, whose lives depend on being sensitive to potential dangers. Because of these natural constraints on sleep, some animals have adapted by developing particularly intense forms of sleep.

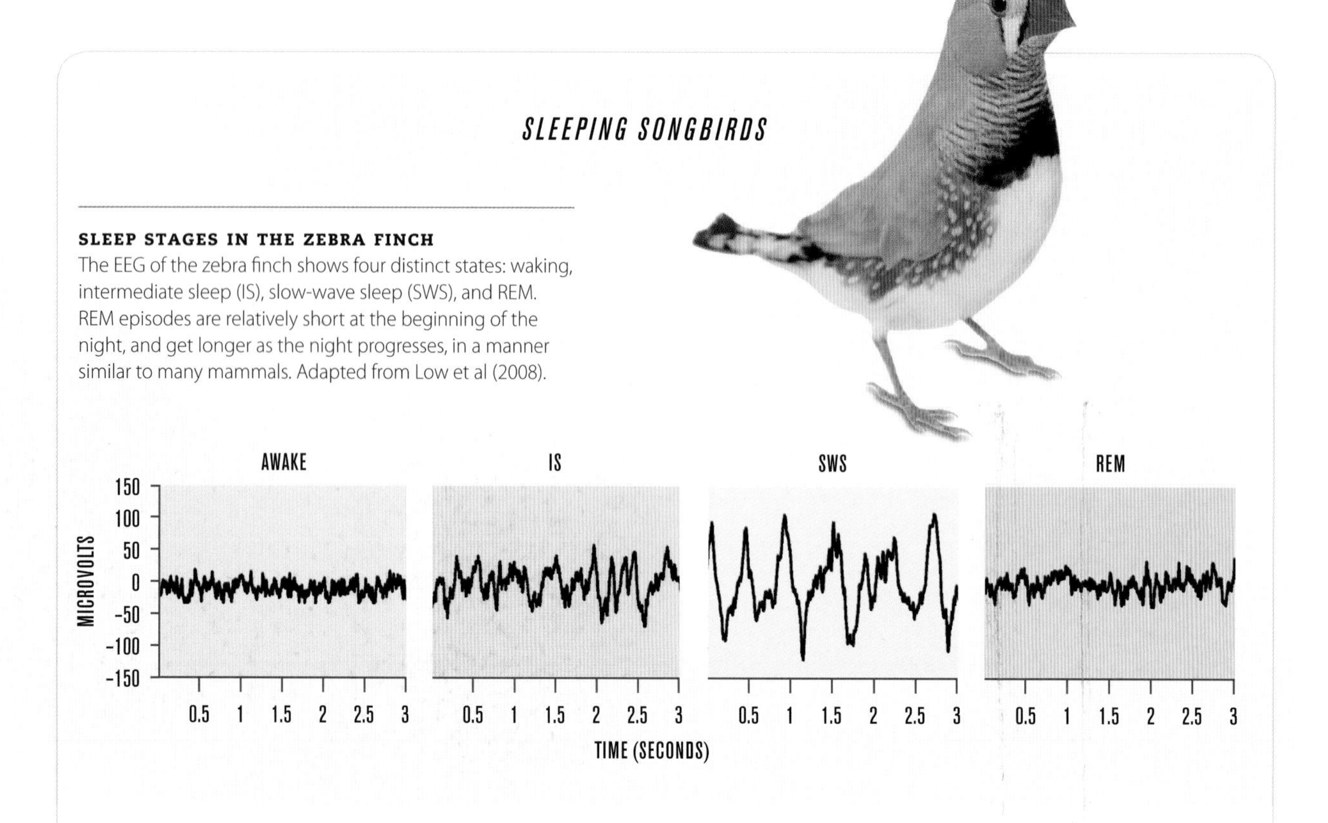

SLEEP STAGES IN THE ZEBRA FINCH
The EEG of the zebra finch shows four distinct states: waking, intermediate sleep (IS), slow-wave sleep (SWS), and REM. REM episodes are relatively short at the beginning of the night, and get longer as the night progresses, in a manner similar to many mammals. Adapted from Low et al (2008).

CETACEANS AND UNIHEMISPHERIC SLEEP

Cetaceans are the only mammals that do not have unequivocally clear episodes of REM sleep. We mentioned earlier the ability of ducks and pigeons to sleep with one half of their brain at a time. This is much more highly developed in dolphins and killer whales, with slow waves appearing alternately on one side or the other (unihemispheric sleep). The bottlenose dolphin, for instance, may sleep on one side of the brain for perhaps two hours at a time, while the other side appears to provide the alertness needed to avoid predators and signal when to surface to obtain air. Cumulatively it may sleep up to 8 hours out of the 24. Interestingly, one can deprive dolphins of sleep on only one side, and the subsequent recovery sleep shows the characteristic deeper patterns of slow wave sleep only on that side. Similarly, giving a sedative such as diazepam (see page 158) can selectively only induce sleep on one side. One particular type of dolphin, the Indus River dolphin, lives in swift currents, where it would be hazardous to be insensitive to the environment. They have evolved a mechanism in which they have many brief "microsleeps," usually of 4–6 seconds duration, while they swim, so that they cumulatively sleep up to 7 hours per day.

In a variation on this theme, fur seals have sleep patterns on both sides of their brain when on land, but when they enter the water they can have slow wave sleep on just one side, with one eye open and one flipper functioning to control their movement. Thus in various species, sleep has developed to meet the needs of the environment in which they are located.

Some scientists believe that, while humans clearly do not have unihemispheric sleep, they may have at least some qualities of hemispheric asymmetry during sleep. A group at the Georgia Institute of Technology examined persons sleeping in an unfamiliar environment, which is usually associated with disturbed sleep (this is known as the "first night effect"). Imaging studies and polysomnography suggested that at different times of the night, sleep depth differed in the two hemispheres, and that the lighter-sleeping side had greater electrical responses to stimuli in the environment. One interpretation of these findings, if replicated, might be that the disturbed sleep in an unfamiliar environment may be a survival mechanism to ensure better responsiveness to potential dangers in a strange place.

The role of sleep in evolution

The role that sleep plays in evolution is still under study. One possibility is that it is an advantageous adaptive state of decreased metabolism for an animal when there are no more pressing activities. This seems true for deeper states of dormancy such as hibernation during the winter when there are few food supplies, and a high metabolic cost to maintaining adequate temperature. It may be true in daily situations as well, for instance for a prey species to avoid predators after dark. On the other hand, the apparent universality of sleep, and the observation that mammals such as cetaceans have developed such elaborate mechanisms to preserve sleep on at least one side of the brain at a time, suggests that sleep additionally provides some vital service(s) for the organism. This is particularly true since one aspect of sleep is decreased responsiveness to the environment. If sleep is ubiquitous even when this potential price must be paid, the implication may be that it has important functions that cannot be obtained just by quiet, wakeful resting. Some of these possible functions are described at the end of Chapter Three (see page 70). Coming back to our earlier topic, it makes it seem less likely that we will find higher organisms that unequivocally do not have sleep or sleep-like behaviors.

DO ANIMALS DREAM WHEN THEY SLEEP?

Sleep and dreaming in animals

This question has been asked from ancient times until the present. Aristotle, in *The History of Animals* concluded that many animals do, including horses, oxen, sheep, and dogs, based largely on the observation of the latter barking in their sleep. In his 1969 novel *Do Androids Dream of Electric Sheep?*, which was the basis of the hit movie *Blade Runner*, the American novelist Philip K. Dick (1928–82) formulated the same question in more modern terms. In an ultimate sense, we can never know if animals have dreams. As discussed in Chapter One, one of the aspects of consciousness (and dreaming would be considered one of the forms of consciousness) is that it is private, that is, not accessible to others. In the practical world, we attribute consciousness to other waking humans in view of their behavior and what they report, and similarly we infer that other humans dream, based on our personal experiences of dreaming and the reports of others. Can we extend this notion to animals? Fortunately, there are ways to examine dream-like behavior and patterns of nerve firing in animals during REM sleep. First, by way of background, we will briefly review some of the qualities of REM sleep.

REM sleep mechanisms

As was discussed in Chapter One, during REM sleep, there is an elaborate neurological mechanism that causes deep relaxation ("atonia") of the major weight-bearing muscles of the body. The purpose of this mechanism is not fully understood. Sigmund Freud, the father of psychoanalysis, believed that the purpose of muscle relaxation during dreaming was to prevent us from acting out their content, which could cause all sorts of difficulties. Indeed, it turns out that there are humans in whom this and related mechanisms are faulty, who engage during sleep in behaviors which can lead to harm to themselves or those with whom they sleep—this is known as "REM sleep behavior disorder" (see page 133). In the 1960s, the French scientists Michel Jouvet and J. F. Delorme discovered that by making lesions in the peri-locus coeruleus (close to the nucleus known as the locus coeruleus—see pages 34–5) in the pons of cats, this atonia mechanism is disrupted. As a result, during this form of REM sleep (known as "REM-A" or REM without atonia), cats engaged in behavior that could be interpreted as acting out their dreams. So, in that sense, what did these lesioned cats dream about? It turns out that they appeared to dream about just what one might guess—they seemed to look about, explore, as well as sneaking up and pouncing on dream mice.

Another way of looking at inferred dreaming is to examine neuronal activity in the brain during REM sleep. In 2007, researchers at the Massachusetts Institute of Technology recorded the firing patterns of neurons in the hippocampus of rats while they were awake and exploring a maze. Not surprisingly, they found that these cells in the hippocampus (an area involved in memory processes) fired with specific patterns that were unique to the physical location in the maze. During REM sleep, they observed the same patterns, implying that during REM the rats were re-creating their explorations. The firing patterns were so specific that it could even be determined where in the maze they appeared to be. In 2015 researchers at University College London took this one step further. Rats were put in a T-shaped maze, which they explored, only to find that at one end was a tempting destination that was out of reach—some

FELINE DREAMS
Lesions in the pons of cats can prevent the normal muscular atonia of REM sleep, and lead to a state know as "REM-A." During REM-A, the sleeping felines engage in behaviors such as exploring or pouncing, which some have interpreted as acting out their dreams.

tasty morsels of food separated by a transparent barrier. As in the previous study, hippocampal neurons during the exploration of the maze fired with specific cells responding to specific locations. During REM sleep, this occurred as well, with the addition of new firing patterns ("pre-play" activation), which could be interpreted as dream explorations of new territory. Indeed, subsequently when awake, the rats were allowed to move into the previously blocked territory, resulting in neuronal firing patterns similar to those during their explorations in sleep. This kind of pre-play activation did not occur for unexplored areas of the maze for which there was little motivation (that is, no food). This seemed to say that rats may dream not only about where they have been, but where they would like to go. It's tempting to draw analogies to human dreams, in which desire often plays significant role. We can never truly know if animals dream, both by the private nature of states of consciousness and because of their limited means of communication, but it is tempting to think that a process analogous to dreams occurs in mammals.

ANIMALS HAVE FAVORITE PLACES TO SLEEP

At night deer often sleep under trees, protected by low-hanging branches, or hidden in tall grass or brush. In the morning, an observant woodsman may see signs of the compressed grass indicating their sleeping spot. Leopards, who sleep mostly in the daytime, find a comfortable tree limb. Not surprisingly, monkeys sleep in the trees as well. Lions, champion predators that they are, can sleep in open areas. Sea otters sleep while floating in the water on their backs. Some birds who sleep in trees are kept safe by a mechanism by which their claws extend deeper into the branch when they are sitting, and retract when they stand up again. Humans have preferred sleeping places as well. When a person spends the night in a new, unfamiliar location, sleep is often disturbed. This is known as the "first night effect."

Chapter Three

THE EFFECTS OF SLEEP DEPRIVATION

Sleep has sometimes been viewed as a "behavioral appendix," a useless process left over from evolution, and an activity for the lazy who should be using their time more productively. We have learned, though, that sleep is necessary for life, and that in extreme cases deprivation of sleep can be fatal in some species. In humans, acute sleep deprivation results in decreased vigilance, as well as alterations in mood and cognition. Many of us live in a state of partial sleep deprivation, without getting all the sleep we need. This leads to deficits in our ability to learn, to drive safely, and for our bodies to chemically process foods that we have eaten, or fight infections. The story is not simple, however. Though most attention has been focused on deficits, some forms of sleep deprivation may be experimentally useful in treating mental illnesses such as depression.

IS SLEEP NECESSARY FOR LIFE?

What happens when we are deprived of sleep?

As we saw in Chapter Two, sleep appears to be a fundamental process to life; sleep-like states appear in even some simple multicellular organisms known as protists and dinoflagellates, and are universal in more complex life forms. What happens, then, when we are deprived of it?

Sleep deprivation is a slightly different topic from insomnia. As we will see in Chapter Eight, insomnia is the personal experience of getting inadequate quantity or quality of sleep or of not feeling refreshed after sleep. In this chapter we will discuss the topic of what happens when people with no sleep disorders are deprived of the sleep they need, due to a variety of reasons such as lifestyle or work obligations. Sleep deprivation can be either acute (short term), or a chronic (long-term) situation of partial deprivation.

For many years sleep was commonly thought of as an idle activity, taking away from the time we could be using to do productive things. A famous proponent of this belief was the American inventor Thomas Edison (1847–1931), who claimed to sleep only four hours a night, often chose employees based on their endurance, and employed watchers to make sure they were not sneaking off to the stairwells to take naps on the job. Such practices resulted in a climate that praised what came to be known as "manly wakefulness." This view was not confined to capitalist, free enterprise societies; at the height of the cold war, the Soviets investigated the possibility that electrical stimulation of the brain might allow workers to get along with fewer hours of sleep and hence be more productive. Some scientists compared sleep to a kind of appendix, a behavioral activity that might have outlasted its usefulness. Although there had been some hints that sleep might be necessary for life in studies of sleep deprivation in dogs in Russian

THOMAS EDISON
The inventor of the light bulb, phonograph, and motion pictures claimed to sleep only four hours per night. A believer that rest was unproductive, he often worked 100-hour weeks, and was known to conduct job interviews at 4:00 AM. In a similar theme, a more contemporary entrepreneur, Jack Dorsey, has said that he spends up to 10 hours a day at each of his two companies, Square and Twitter.

CHARLES LINDBERGH

As the American aviator Charles Lindbergh (1902–74) prepared for his New York to Paris flight, he planned for all sorts of eventualities. He was so obsessed with minimizing weight that he dispensed with a radio and substituted an uncomfortable wicker chair instead of the usual heavier leather. A mail pilot with many overnight trips behind him, he felt that he could manage the fatigue. The takeoff had been delayed by bad weather with seemingly no prospect for relief, so on a rainy Thursday May 19, 1927, he was persuaded to go into Manhattan to see the Broadway show "Rio Rita." While there, one of his team checked with their meteorologist, who predicted (not altogether accurately) clearing skies over the Atlantic. Lindbergh rushed back to Roosevelt Field, spent several hours preparing the plane, and went to bed around midnight with orders to get him up at 2:15 AM. After drifting off to sleep, he was awakened by an aide with an unnecessary question, and was unable to go back to sleep, arising at 1:40 AM. Having essentially been up for 23 hours, he decided to fly anyway, and lifted off at 7:51 AM. During the next 33.5 hours, often flying only by instruments in a freezing weather front, he became progressively more sleepy. He is said to have used his thumbs to prop his eyelids open. At one point he resorted to flying so low that the cold ocean spray would hit his face through the side windows. He began to hallucinate, and later described ghosts who joined him in the cockpit. Finally he crossed the coast of Normandy and followed the Seine to Paris. After being awake for 56 hours, with no runway lights or ground communications to help him, he successfully landed.

studies going back as far as 1894, this idea never gained traction until the 1980s, when studies at the University of Chicago found that two to three weeks of sleep deprivation was fatal to rats. When they were sleep deprived for shorter times, they survived, and had long periods of recovery sleep in which REM sleep was greatly elevated. The rats were also selectively deprived of REM sleep, which turned out to be fatal after about four to six weeks. These studies were dramatic demonstrations that sleep is necessary for life. Moreover, during the sleep deprivation period the animals had alterations in weight, appetite, and temperature, a reminder that some of the areas in which sleep is regulated, such as the hypothalamus, and some of the neurotransmitters involved in the control of sleep, such as serotonin and norepinephrine (noradrenaline), are involved in a variety of other physiological processes.

ACUTE SLEEP DEPRIVATION

Decreased alertness and performance after acute sleep loss

After missing a single night's sleep, most of us will experience the sensation of sleepiness, and have the feeling that performance declines. Interestingly enough, after this very short duration of deprivation, one will indeed fall asleep sooner if given an opportunity to nap, but objective deficits in performance may be relatively mild and reversible after only a few hours of sleep. Indeed, by some measures they might even slightly improve, presumably because one is trying harder. As duration of sleep deprivation gets longer, sleepiness is more evident both when measured subjectively by questionnaires such as the Stanford Sleepiness Scale, or objectively by the Multiple Sleep Latency Test (MSLT), in which a person is given four opportunities to nap across the course of the day, and the average time to fall asleep (sleep latency) is measured (see pages 128–9). Behavioral changes include alterations in mood, alertness, and performance. Mood changes are among the first features noticed with sleep deprivation, with increases in fatigue and tension, and a decline in vigor on a standardized test (Profile of Moods, or POMS). The American sleep researcher William Dement (1928–), then a medical student who assisted in the Peter Tripp study mentioned opposite (later going on to have a very distinguished leadership role in sleep research) described the effects of staying up for 48 hours while assisting in research in Nathaniel Kleitman's laboratory. He began to feel that his room-mates were mad at him and plotting against him, even though he realized at the same time that suspiciousness could be part of sleep deprivation. These feelings persisted until he had an opportunity for recovery sleep. Vigilance and alertness can be assessed by standardized procedures in which a person is given a not very demanding repetitive task and determining number of errors across time. Performance measures include tests in which subjects are asked to remember lists of words or convert numbers and symbols. Performance tests are usually initially less sensitive to sleep deprivation than measures of mood and alertness. In general the kinds of tasks that are most impaired are those that are more complex, and involve speed and learning.

Decreased performance

There are many influences on the degree to which sleep loss results in deficits. Among these are the time of day or night, the amount of time awake before sleep, and the duration of the preceding sleep. (When the length of the preceding sleep period is shorter, the effect of subsequent sleep deprivation is more pronounced.) During the deprivation period, arousal can be influenced by many factors, including physical activity. After relatively mild sleep deprivation, perhaps missing half of a night's sleep, even a 5-minute walk can improve wakefulness, though this effect declines with significant sleep loss of roughly 40 or more hours. Being in a brightly lit area improves wakefulness during sleep deprivation. Though music may have some alerting qualities under normal conditions, its benefits become quite small after sleep deprivation. Background noise, which can impair performance when one is alert, may have a small beneficial effect during sleep deprivation, presumably by increasing the level of arousal. There is little evidence that changing the temperature improves wakefulness after sleep deprivation. Older people respond about the same to sleep deprivation as younger individuals, and

PETER TRIPP'S SLEEPLESS MARATHON

In 1959, Peter Tripp (1926–2000), an American Top 40 countdown DJ for WMGM in New York City, agreed to help the March of Dimes by staying awake for 201 hours while broadcasting from a glass booth in Times Square. For the first several days he seemed relatively intact though progressively irritable and rude. The story goes that he was so insulting to his barber of 20 years that the poor man went away in tears. Starting around the fourth day, Tripp began to see things that weren't there, cobwebs on his shoes, and strange cats that no one else could see. He became very suspicious, particularly at night, thinking for instance that the technician helping him was actually planning on burying him, and that enemies were putting medicines into his food to make him sleep. He experienced auditory hallucinations. Ultimately he believed that he was not Peter Tripp. He began feeling colder as his temperature dropped, and asked for more clothing while others felt comfortable. In the end he was unable to recognize his wife. When the experiment finished, he had a long sleep, and said that he felt fine. Whether he was or not has never been clear. His wife divorced him, which some have taken to be evidence that something went wrong, but it is noted that all four of his marriages ended in divorce. A few weeks after the experiment, he was indicted in one of the first of what came to be known as the payola scandals, allegations that disk jockeys were taking money from record companies to play certain songs on the air. His career never really recovered.

there is a little evidence that older men preserve performance somewhat better than their younger counterparts during the night. Obviously, motivation influences the degree of performance during sleep deprivation, and there is even evidence that knowing the deprivation will end soon can be helpful.

There is also a significant difference among individuals in sensitivity to sleep loss, due to qualities which are poorly understood but which may be related to differing levels of brain activation before sleep loss. Genetic studies have associated various forms of the gene known as PER3 with different degrees of subjective sleepiness and memory function during partial sleep deprivation followed by total sleep deprivation. It may be that people who are extroverts and those who are more sensitive to caffeine are more affected by sleep deprivation. Nicotine seems to have no significant effect on wakefulness or performance, at least during studies of one or two days of sleep deprivation.

Acute sleep loss: changes in memory and physiologic processes

Acute sleep loss may decrease the ability to store memory of materials learned during the day into the long term, as we will discuss in more detail in the section on partial sleep deprivation (see pages 60–3). There is a much greater sensitivity to sedating effects of alcohol in a sleep-deprived person, as demonstrated in simulating driving tasks, which resemble video driving games (and sadly, in the real world). As we saw in the story of Peter Tripp, many people will develop mild mood changes including depression and anxiety, and some paranoid thoughts during sleep deprivation. Misperceptions and visual hallucinations may occur and, after prolonged deprivation over 100 hours, a small percentage of individuals (2 percent in one study) may develop significant psychotic states, usually with prominent paranoid features. This is more likely to appear in people with a psychiatric history. These states improve rapidly following the end of deprivation.

A number of physiologic changes are also found. After 24 hours of sleep deprivation, there is an increase in diastolic blood pressure due to alterations in autonomic nervous system function. In terms of the EEG, after one to several days of sleep loss there is a progressive decline in alpha activity or ability to produce alpha upon closing the eyes (see pages 12–13). Delta and theta activity increase in the EEG, and there may be brief periods of EEG slowing which are considered to be very brief ("microsleep") episodes. Neuroimaging studies indicate that after one night of sleep loss, there is reduced activation of parts of the brain known as the fronto-parietal attention network and the salience network, biological measures consistent with the decreased ability to maintain attention and vigilance.

People who have been deprived of sleep for two nights have also been found to have impaired ability to make decisions in uncertain situations which require dealing with incoming information or unexpected events. In principle, this might be of concern in disaster relief or military situations. Seizures have been reported in non-epileptic soldiers who have been awake for more than 24 hours, and occasionally EEG spiking (a sign of potential seizures) is seen in sleep-deprived people. This happened, for instance, in American servicemen returning from Vietnam. They generally had been up all night celebrating the night before they left, then were awake for a long trip that often took more than 24 hours. After landing in America, they might have a seizure, never to have another the rest of their lives. In a more contemporary setting, this same phenomenon has been reported among young employees of Wall Street financial firms. One example was a first year analyst at Goldman Sachs who was hospitalized following a seizure, which occurred after he had worked continuously for 72 hours. Another, a 21-year-old intern at Bank of America Merrill Lynch, had a seizure and died while taking a shower before going to the office, after having worked 72 hours without sleep.

Other neurologic changes are relatively minor. After very prolonged sleep loss of 200 hours or more, a subject may have slurring of speech, tremors, and abnormal eye movements known as nystagmus. Pain sensitivity is increased (see page 67), and animal studies suggest that it may not return to normal for up to a day after the end of deprivation. Acute sleep deprivation produces only minor changes in maximal exercise ability in humans, although it may slow down the rate of recovery from exercise. There are minor changes in insulin and blood sugar regulation after acute sleep deprivation; these become very significant after chronic partial sleep deprivation, as we will see below.

One interesting ongoing area of research is whether medications might ameliorate the effects of acute sleep loss. A study of sodium oxybate (which is discussed in Chapter Seven as a treatment for narcolepsy) found that after a night of sleep

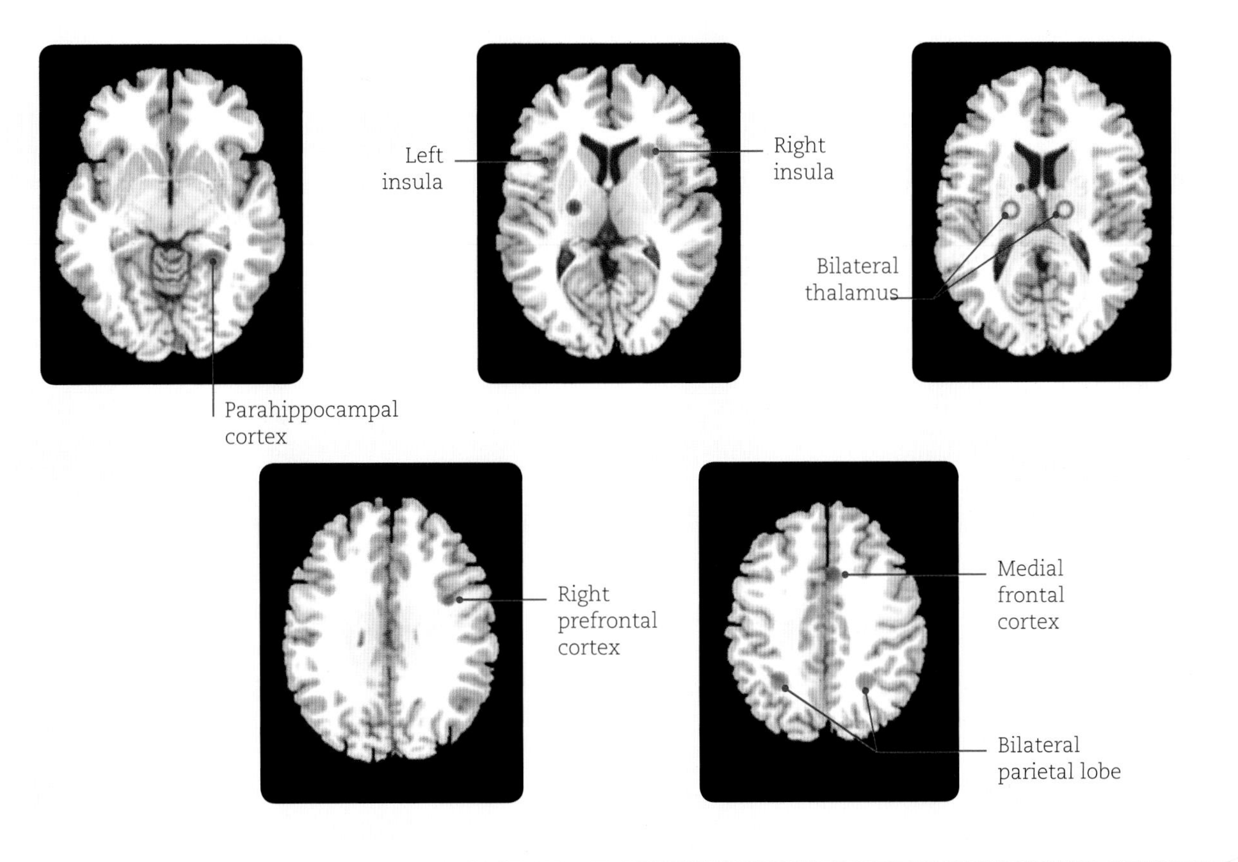

deprivation, it improved performance on the Psychomotor Vigilance Task, and reduced sleepiness on a subsequent Multiple Sleep Latency Test, or MSLT (see pages 128–9), while increasing slow-wave sleep during recovery sleep. When given during four nights of sleep restriction, the anticonvulsant drug tiagabine has been reported to increase slow-wave sleep and improve sustained attention the next day. One study indicated that the investigational drug gaboxadol, which stimulates receptors for the inhibitory neurotransmitter GABA (see pages 36–7 and 158–9), increases slow-wave sleep and subsequent wakefulness during four days of sleep restriction.

NEUROIMAGING STUDIES

These pictures represent a composite of data from 11 studies using functional magnetic resonance imaging (fMRI), which measures local blood flow in the brain as a reflection of neuronal activity. When acutely sleep-deprived subjects engaged in tasks requiring attention, a variety of brain areas showed decreased neuronal activation (blue). Among these are the frontal and parietal cortical areas, which are important in attentiveness, and the insula, which promotes attention and connection to working memory when a relevant stimulus appears. Some areas (yellow and red) display heightened activity. Among these is the thalamus, which is important in arousal and alertness. It has been speculated that perhaps increased activity in the thalamus reflects a greater effort to compensate for decreased function in the frontal and parietal regions, or that greater thalamic activity is required for wakefulness during attention tasks following sleep deprivation. Adapted from N. Ma et al (2015).

PARTIAL SLEEP DEPRIVATION

What happens when people are limited in the amount of sleep they can have?

It's pretty clear that many of us sleep substantially less than we need to. A large 2008 poll known as "Sleep in America" found that working adults estimated that they needed an average of 7 hours and 18 minutes of sleep in order to be at their best, but 44 percent said they usually slept less than 7 hours on weeknights. In the United Kingdom, 65 percent of the adult population describe sleeping only about 6.5 hours per night.

There is some issue as to whether sleep is becoming progressively shorter in modern society. A study by the National Sleep Foundation is typical of reports that this might be the case, indicating that in 2009, 20 percent of Americans said they were sleeping less than 6 hours per night, compared to 12 percent in 1998. Other studies have suggested that this is less clear, and may be dependent on how the question is phrased. In a study in which sleep duration was derived from time studies of how people spend their whole 24 hours, rather than just asking them how much they sleep, the change in percentage of people reporting sleep of less than 6 hours was very small between 1975 and 2006, and significant only for full time workers, those with college education, and African-Americans. Another approach is to look at objective data from the sleep laboratory. One review of such studies comparing the periods 1960–89 and 1990–2013 found no difference in objectively measured total sleep time.

Sleep across the world

Habitual sleep durations also differ across the world. Studies of sleep in students indicate that the Japanese get the least sleep (a little over 6 hours), while Bulgarians get the longest (close to 8 hours). Studies in Finland and Australia have suggested that there may also be genetic components to habitual duration of sleep. There has been some speculation as to whether different sleep durations may play a role in overall health differences between races, though there are so many factors to be considered that it is uncertain whether the relation is just associational or causal. Income status also plays a small role in reported sleep duration. The National Health Interview Survey by the US Centers for Disease Control in 2013 found that a higher percentage (35.2 percent) of people living below the statistical poverty line slept less than 6 hours per night, compared to 27.7 percent of those whose income was greater than four times that amount.

JAPANESE SLEEPING HABITS
As seen in the graph opposite, the Japanese have among the shorter reported nocturnal sleep times. To some degree, this may be encouraged by praise, for instance for businessmen who get little sleep in order to maximize work. Possibly in compensation for shorter sleep, there is also more acceptance of "inemuri," the practice of sleeping briefly in the daytime in public places, including commuting trains or parks.

SLEEP DURATION AROUND THE WORLD
Reported sleep durations in various countries, as determined in a time use study by the French-based Organization for Economic Co-operation and Development (OECD) and related sources. Studies can differ in results depending on how the question is asked, for instance whether the emphasis is on how people report dividing their time across the 24 hours, or whether they are specifically asked about sleep habits.

Effects of limited sleep on health

Although there is some issue as to whether amounts of sleep are actually declining, the significant number of people who currently report relatively short periods of sleep makes it important to consider what this might mean for health. In Chapter One we made reference to studies associating a higher mortality (death rate) with sleeping substantially less or more than 7 hours, as well as a variety of illnesses associated with inadequate sleep. We will focus here on the portion of people who on average sleep less than 7 hours per night on a long-term basis. Prior to the late 1990s, studies (often not done in the laboratory) seemed to indicate that sleeping 4–6 hours per night for months at a time caused little deficit in alertness or performance or feeling sleepy. Starting around that time, more sophisticated laboratory studies began to come to very different conclusions. Firstly, it is now recognized that, in contrast to acute sleep deprivation, in chronic partial deprivation, a person often does not have the sensation that (s)he is sleepy. This can be alarming, as a person who does not recognize (s)he is sleepy may be more careless about safety than a person who realizes (s)he is impaired. In Chapter Eight we will see that this same concern comes up with chronic use of some longer acting sleeping pills, which can impair daytime performance without a person being aware of it. Secondly, a large body of evidence of various types of performance deficits, as well as long-term physiological changes, is now recognized.

As in the case of total sleep deprivation, there is also a significant difference between individuals in their ability to tolerate partial sleep deprivation, for reasons that are not yet clear. The most obvious effect of partial sleep deprivation is, of course, sleepiness. In the preceding section we mentioned that among the factors affecting the degree of arousal during sleep

HOW MUCH SLEEP DO WE NEED?
A 2015 consensus statement from experts in many fields, assembled by the American organization the National Sleep Foundation, made these recommendations. For many age groups the recommended sleep times were widened, reflecting the growing recognition of individual variability in sleep need.

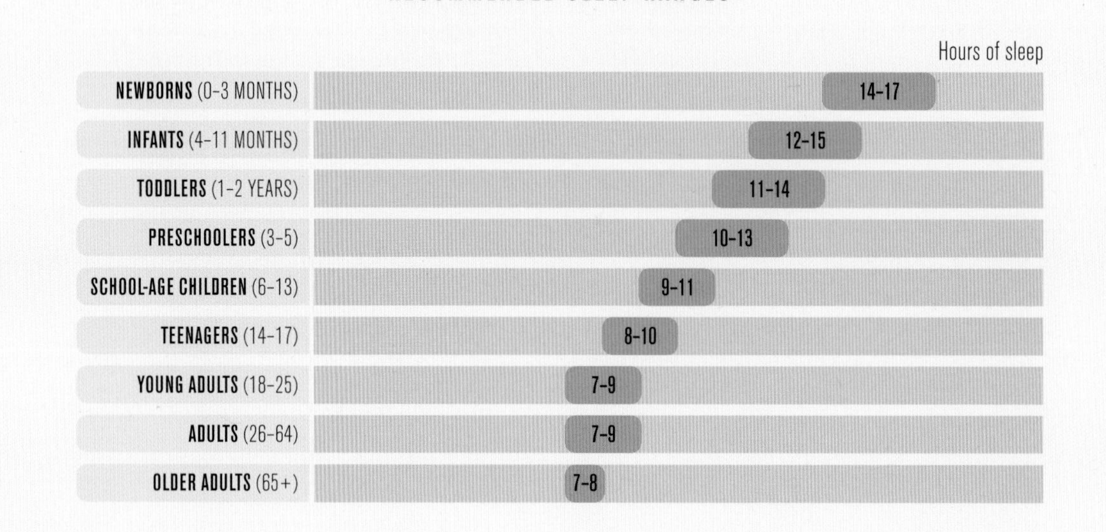

deprivation is the duration of the prior sleep period. This is evident, and cumulative, during partial sleep deprivation. For instance, in people whose sleep has been reduced from 7 to 5 hours nightly for one week, daytime sleepiness as measured objectively on the MSLT is increased on the first day, then becomes progressively greater on succeeding days. People whose sleep is restricted by 40 percent over five nights show significant impairment in vigilance and simple reaction time. A commonly used measure in these types of studies is the Psychomotor Vigilance Task (PVT). In this procedure, the subject is asked to press a button whenever a light randomly appears on a screen every few seconds. The number of times (s)he fails to push the button is taken as a measure of lapses in vigilance. In one study of young adults restricted to 4 hours' sleep per night, after two weeks their performance on the PVT test was as impaired as subjects who had had no sleep at all for one or two nights. Driving simulator studies show increased rates of accidents, even when sleep is only restricted to 7 hours, and significantly more when reduced to 4–6 hours. As we mentioned earlier, one of the biggest concerns is that in these studies, subjects may become extremely impaired, but believe that they are only moderately sleepy, and hence not be good judges of the degree of their impairment.

Sleep is also important in converting information learned during the day into long-term memory (a process known as "consolidation"). This is true for a variety of learned material including motor skills, mathematics, and tests of word association. Initial studies suggested that declarative memory (consciously recognized facts and verbal knowledge) may be more related to REM sleep, while procedural memory (memory of how to perform a particular action) may be more associated with slow-wave sleep. (As we will see later in the section on REM deprivation, this seems less clear now than it once did.) Children and adults whose total sleep has been experimentally limited are less effective the next day at retaining some kinds of previous learned material.

EFFECTS OF SLEEP SCHEDULES AND CAFFEINE ON PERFORMANCE

It is becoming possible to predict how a particular sleep–wake schedule and amount of caffeine consumption can affect alertness and performance in groups of normal individuals. At the web site 2B-Alert Web (2b-alert-web.bhsai.org/) one can see how inputting information about the amount and timing of sleep will likely affect measures such as response time and lapses in performance. This can show how groups of normal individuals will react on average to a certain schedule, and is not predictive of how a single individual will respond, but it can be useful in seeing the general effects of different amounts and timing of sleep.

SLEEPING IN SPACE

Astronauts on space shuttle flights have markedly reduced sleep, not only during the mission but in the months of training before the flight. Many factors go into this, including the excitement of the mission, the rapidly changing light-dark schedule (the sun rises and sets every 90 minutes in these situations), and the very significant noise levels in spacecraft. One suggestion, that they wear earplugs, was often not received with enthusiasm. It turns out that astronauts generally preferred not to, as changes in ambient sound could often be the first sign of danger. Similar remarks have been made by some long-haul airline pilots, who are given scheduled opportunities to take naps during long flights. They have said that one thing that brings them out of their sleep immediately is the sound of the engines changing speed. The effects of maintaining this degree of attentiveness to the environment are significant. It even turns out that the situation of being on call without ever having been awakened—as in a study of ship engineers—disrupts sleep.

Physiological changes

A variety of physiological changes occur in people who are sleep restricted. There are alterations in the immune response, including levels of "natural killer cells" and cytokines (chemicals related to the immune response). In one study, after a single night in which healthy subjects were kept awake until 3:00 AM, the proportion of natural killer cells dropped, as did the amount of an important cytokine. Although it is always difficult to translate these physiological measures into practical outcomes, it has been reported that people who sleep less than 5–6 hours per night are four times as likely to catch colds as those who sleep at least 7 hours. As with acute total sleep deprivation, there is also an increase in pain sensitivity. Normal subjects who are 50 percent sleep deprived for two nights, for instance, have greater subjective and EEG response to moderately painful electrical stimuli to the forearm, and a lower pain threshold for pressure pain to the shoulder area. As little as two nights of sleep restriction (time in bed from 2:45 to 7:00 AM) has been shown to alter bacteria in the intestines, although it has not been established whether this relates to the various metabolic changes which result from limited sleep.

Heart attacks and related events are more frequent in nurses who sleep less than 7 hours compared to those who sleep 8 hours. There is some reason to think that this connection is related to increased inflammatory activity, as measured for instance by the chemical blood marker C-reactive protein. Population studies have also shown that chronic partial sleep deprivation is a significant risk factor for developing hypertension.

One of the most interesting changes with sleep restriction is an increase in appetite. During sleep restriction in animals, activity of the orexin/hypocretin system (peptides in the lateral hypothalamus thought to be related to both arousal and feeding) increases, and the amount of hypocretin-1 in the cerebrospinal fluid rises. Orexin/hypocretin release is in turn affected by other hormones that regulate appetite. Grehlin, a hormone that stimulates appetite, is increased, while leptin, a hormone which gives a sense of satiety, declines. In humans as well, an increase in appetite after partial sleep deprivation has been demonstrated. A review of 11 studies found that after sleep was restricted to about 4 hours, people tended to consume about 385 more calories the next day. Studies of sleep-restricted individuals show that in the morning they are unable to utilize the sugar glucose as effectively and may have a diminished ability to secrete insulin when needed to regulate blood glucose. An MRI (magnetic resonance imaging) study indicates that after sleep restriction, there is increased activation at brain centers responsive to food stimuli, as well as reward centers. Short sleep is also associated with increased appetite, more difficulty in losing fat during dieting, and increased late-night eating in some individuals. There have been many studies relating sleep restriction to increased body mass index (BMI). Several studies that followed subjects for a decade or more found that chronic short sleep was associated with an increased risk of diabetes.

In terms of brain physiology, we mentioned in Chapter One that one of the functions of sleep may be related to cleaning out waste products, through a modified lymphatic system known as the "glymphatic system." It has been speculated that inadequate amounts of sleep may result in a chronic deficit in the ability of the nervous system to remove

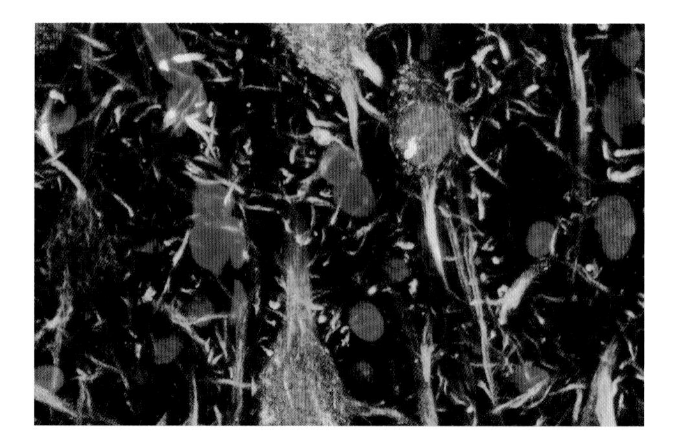

OREXIN RECEPTORS IN THE BRAIN
The orexin/hypocretin system, centered in the lateral hypothalamus , which promotes both wakefulness and food intake, is stimulated by limiting sleep. In this photograph taken using a specialized technique known as confocal light microscopy, receptors for orexins are seen in red, while neurons are pictured as green.

EFFECTS OF LIMITED SLEEP ON THE BODY

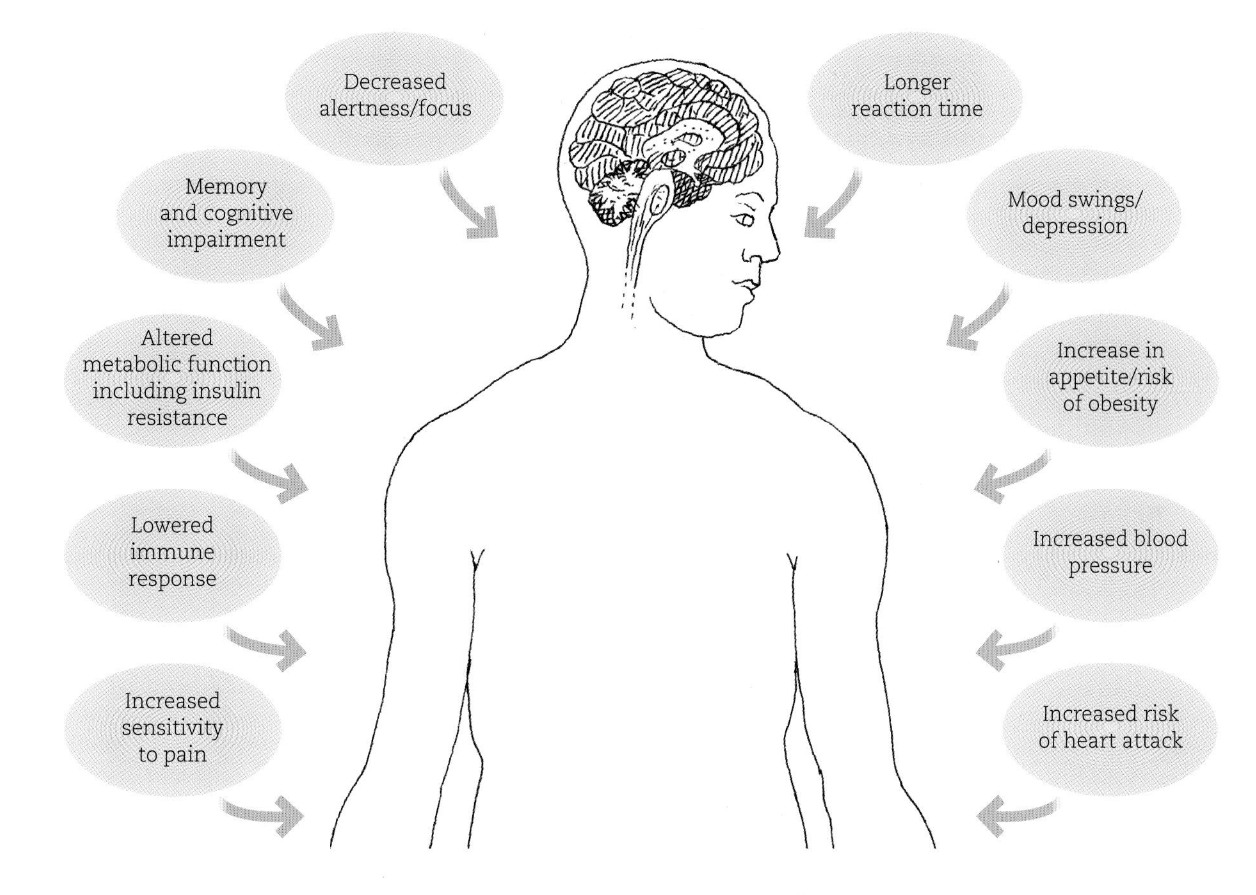

HEALTH EFFECTS OF CHRONICALLY LIMITED SLEEP
Limiting sleep to less than what is needed can lead to risk for a variety of physiological changes. One of the most striking features is the wide range of organ systems that are affected by, and in turn influence, sleep.

potentially toxic waste materials. Since degenerative illnesses such as Alzheimer's and Parkinson's disease involve accumulation of abnormally folded proteins, it is possible that inadequate sleep plays a role in their genesis. Disturbed night-time sleep and daytime sleepiness in older people are among the risk factors for later developing Alzheimer's disease. The possible relation of sleep to these types of disorders is still being explored.

The practical implications of chronic partial sleep loss are significant in a variety of settings. Medical residents have been found to have increased rate of medication errors and accidents after being on call. It turns out that about half of medical trainees describe having fallen asleep while driving, particularly after being on call. Nurses who work rotating shift schedules and develop sleep difficulties have higher rates of accidental needle sticks. Police working night shifts have higher injury rates. In all these cases, however, it is hard to separate the roles of chronic sleep deprivation and that of circadian (body clock) rhythm effects (see Chapter Four, pages 72–85). The interaction of sleep loss and circadian effects is such that sleepiness is greater at night, degrading performance especially when working for prolonged periods ("time on task"). Commercial truck drivers face many of these difficulties, often compounded by an increased rate of obstructive sleep apnea, which of course also causes sleepiness. This is made more difficult yet by observations that sleepiness results in a greater tendency toward impulsive behavior and risk-taking.

SELECTIVE SLEEP DEPRIVATION

What happens when a person is deprived of a specific stage of sleep?

REM sleep deprivation

The effects of REM deprivation are complex. REM has been hypothesized to have an important role in memory, and initial animal deprivation studies seemed to show deficits in converting newly acquired information into long-term memory. Later studies have been negative, and retrospectively it seems possible that many of the claimed deficits were related not to REM deprivation per se, but rather than to the stress of the study. The original view that in humans REM might be crucial to consolidation of some kinds of memory now seems less clear.

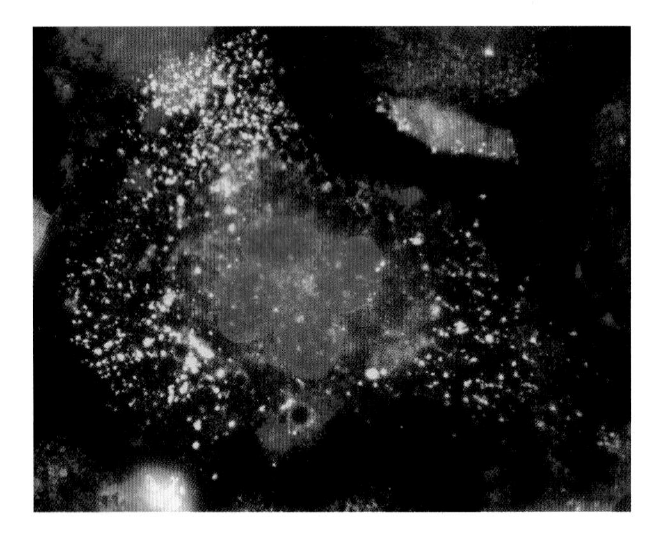

CYTOKINES

Macrophages, a type of white blood cell which produces cytokines, as seen using a technique known as immunofluorescent light micrography. The nuclei of the cells are in blue. The yellow dots represent locations of a protein which modulates the amount of cytokines which are made. In rats, REM deprivation leads to higher circulating levels of pro-inflammatory cytokines, which might affect both immune and metabolic processes. A macrophage visualized with a different technique known as scanning electron microscopy can be seen on page 141.

Certainly one strike against it is the observation that many people take medicines which greatly suppress REM sleep (such as a kind of antidepressant known as monoamine oxidase inhibitors) for years, without reporting problems with memory. This is similarly true of patients with damage to the area of the brain stem known as the pons (see pages 34–5), who live quite successfully without REM sleep. In one well-publicized case, such an individual went on to finish law school, and later became a newspaper puzzle editor. A more recent study of patients who take REM-reducing selective serotonin reuptake inhibitors (SSRIs) found no deficits on formal memory testing, and on one particular memory task (motor learning) actually found some improvement.

In general, it has been thought that the role of REM is particularly important in consolidation of more emotional memories, although at least one study in normal young adult volunteers did not find this effect. In this particular study, they were shown emotionally neutral and negative pictures. Recognition of pictures was better for negative than neutral ones, and improved after subsequent sleep. Those who were REM deprived, however, did just as well as those who had been allowed to sleep through the night.

The effects of REM deprivation in animal studies are various. In some cases, animals have increased drive behaviors, with increased motor activity, sexual behavior, and eating. If done after a spatial learning procedure in rats, subsequent REM deprivation has been reported in some studies to impair consolidation of the newly learned material, although others have not found this effect. Following deprivation, liver enzymes may become abnormal, and there may be signs of inflammation as evidenced by the rise of cytokines such as some of the interleukins. An effect on the immune system is also suggested by a study indicating that mice who are REM deprived are less likely to survive malaria infection. Brain concentrations of norepinephrine, and overall excitability increase.

Slow-wave sleep deprivation

Deprivation of slow-wave sleep is more difficult to perform, as it requires over five times more awakenings of the subject than REM deprivation. Slow-wave sleep deprivation for three nights in normal female volunteers produced fatigue, a greater sensitivity to pain, and an increase in an inflammatory skin flare response. These are features sometimes seen in the disorder known as fibromyalgia syndrome, characterized by fatigue and muscle pains, suggesting to some that a deficit in slow-wave sleep may play a role in causing it. Another study, which was examining the disruptive effects of periodically disturbing sleep every 10 minutes throughout the night, found sleepiness and performance deficits the next day; when slow-wave sleep specifically was prevented by awakenings, there was no increase in sleepiness, or mood or performance difficulties. This seemed to suggest that frequent disturbance of sleep across the night, rather than specifically a loss of slow-wave sleep, results in a loss of the restorative quality of sleep. There is some evidence that slow-wave sleep deprivation without an alteration in total sleep time may acutely improve symptoms in depressed patients without a change in vigilance.

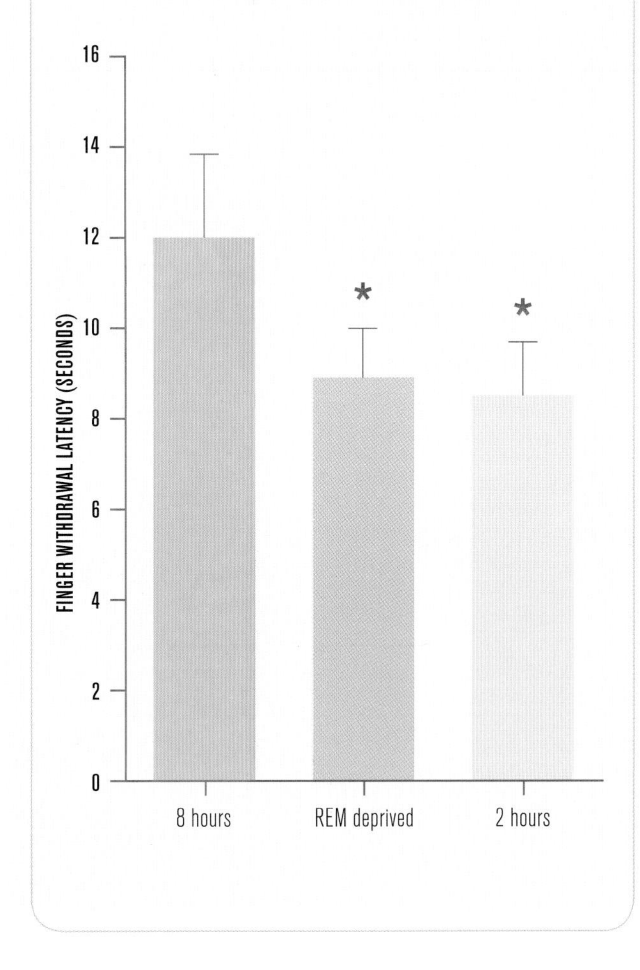

PAIN RESISTANCE AND SLEEP DEPRIVATION
In this study, pain sensitivity in volunteers was measured by determining how quickly they removed their fingers (finger withdrawal latency) from a spot receiving heat from a 100-watt projection light bulb. The shorter time duration indicates greater pain sensitivity. After two nights of reduced total sleep or selective REM sleep deprivation, they were found to be significantly more sensitive to pain. The asterisks indicate statistical significance, in this case that there is a less than two percent chance that the difference in results between these deprivation groups and the 8-hour time-in-bed group is due to random chance. Adapted from T. Roehrs et al (2006).

IS SLEEP DEPRIVATION ALWAYS BAD?

Can sleep deprivation be beneficial in certain situations?

The general assumption has been that decreased sleep is harmful in various ways which we have described earlier, but a few studies such as the last one we mentioned suggest that, in certain circumstances, decreased sleep or reduction in particular sleep stages may be beneficial. We will look at a few of these beginning with REM deprivation as a possible treatment for depression.

In the 1960s, scientists at the US National Institute of Mental Health noted that depressed patients often initially had decreased sleep in general and in particular disruption of REM sleep. The thinking at the time was that REM sleep deprivation might result in mental disorders. Thus it was speculated that one road to depression might be that a person has had something bad happen and as a result sleeps poorly; the poor sleep results in decreased REM, which in turn results in depression. This overall notion has largely been discarded for a variety of reasons, including the later recognition that REM deprivation is unlikely to be harmful to mental health, as well as the finding that in depression REM sleep often appears earlier at night and for longer initial durations, and that often in a developing depression sleep disruption may not occur until very shortly before mood symptoms.

REM deprivation and depression

In the 1970s, Gerald Vogel (1928–2012), an American psychiatrist at Emory University, turned this idea on its head. Noting that the main antidepressants of the time (tricyclic antidepressants and monoamine oxidase inhibitors) were powerful suppressors of REM sleep, that electroconvulsive therapy usually reduced REM, and that in animal studies REM deprivation increased drive behaviors such that they ate more food and became hypersexual, he wondered if REM deprivation might be a treatment for depression. He put depressed patients in the laboratory and awakened them every time they entered REM sleep for up to 30 forced awakenings or six days, and repeated this procedure for several weeks. It turned out that over a few weeks their depression improved, to the same degree as those treated with the antidepressant imipramine, and that after cessation of treatment, depression did not return any more rapidly than expected from controls. Obviously, this was not a very practical treatment because of the intensity of the labor involved, but it raised the possibility that changes in sleep may not just be a consequence of depression but might somehow be closely involved in its genesis.

Total sleep deprivation in depression

In the 1980s a number of studies found that total sleep deprivation also might improve depressive symptoms in up to half the patients as quickly as the afternoon after deprivation. It even turned out that just keeping patients up for half the night was also therapeutic. Deprivation of the first half or second half of sleep had equal effectiveness, as long as the amount of REM sleep remaining was the same. One thing that the total and partial sleep deprivation studies have in common, however, is that though the antidepressant response occurs, often after just one night of deprivation, its benefits usually disappear after the first recovery sleep. Indeed, some patients became depressed again after just taking a nap the next day. Again, these studies have been important in asserting the close relationship between sleep and depression, but have limited practicality in

SLEEP DEPRIVATION IN RESPONSE TO TRAUMA

There is reason to think that sleep deprivation may have an interesting effect on the way upsetting events are dealt with. In one study, normal volunteers were shown a traumatically themed movie as an analogue of a traumatic experience; afterward some were allowed to sleep as usual while the others were kept up all night. The next day, the sleep-deprived group were less affected by the upsetting experience of seeing the movie, and had less intrusive thoughts about the movie over the next six days. In this case it is not clear why sleep deprivation seemed to reduce the impact of the upsetting event. One possibility is that since sleep seems to be involved in memory processes, perhaps the absence of sleep during the deprivation night helped prevent the upsetting event take a more permanent hold in memory. Statistical analysis indicates that there is a less than five percent probability that the difference in results between the two groups is due to random chance. Adapted from Porcheret et al (2015).

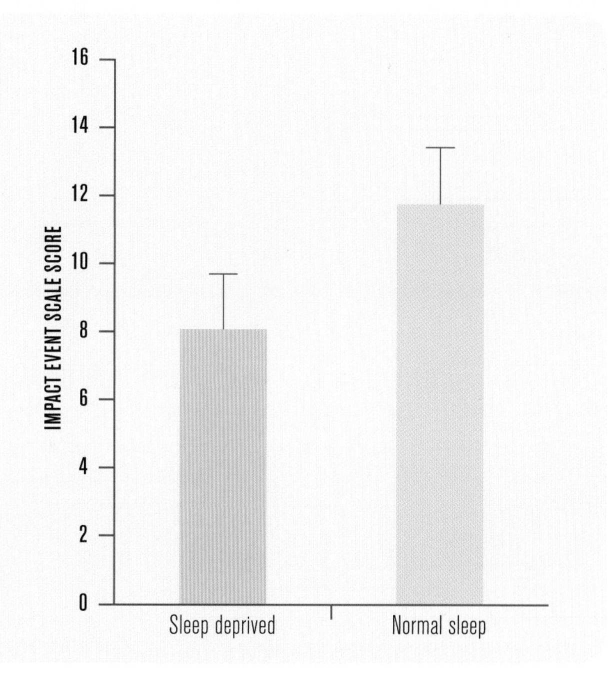

application. There was some evidence that sleep deprivation in combination with antidepressants potentiated the benefits of the medication, but overall this remains uncertain. There is also the concern that in patients with bipolar illness, sleep deprivation might induce a manic episode.

A reduction in amounts of slow-wave sleep is typical in depressed patients, which has led some researchers in the last few years to consider whether some disturbance of slow-wave sleep, or perhaps the homeostatic mechanism (for which the amount of electronically measured slow waves, known as delta activity, is often used as a proxy) might play a role in depression. As we mentioned earlier, there are now data that, after one night of slow-wave sleep deprivation, in which slow wave sleep is disrupted by electronic tones without awakening the subject, about half of depressed patients improve symptoms the next day, without evidence of decreased vigilance.

Do these various studies suggest that overall it is better to have less sleep? Not likely, at least as a generality. We have already looked over the many kinds of studies that indicate that chronic short sleep is associated with a number of health concerns. But it does raise the tantalizing possibility that sleep regulation may be intimately involved in the physiology of some psychiatric illnesses.

Experimentally altering sleep can improve symptoms, in a way that is not very practical, but that gives us new ways of looking at the illnesses.

Recovery sleep

There are several interesting qualities about recovery sleep following sleep deprivation. For one thing, the amount of recovery sleep is often much less than the amount that had previously been lost. Even after very prolonged sleep loss of over 200 hours, young adults typically sleep only 12–15 hours. The amount of recovery sleep, the rate of return of normal wakefulness and performance, and the EEG during sleep are influenced not only by the duration of prior deprivation but also by the time of day or night. Another feature is that different deficits such as sleepiness and performance may recover at different rates. In one study of young adults kept awake for 1–2 nights, subjective reports of sleepiness and response speed returned to normal after one 9-hour recovery sleep. An objective measure of sleepiness (sleep latency) did not return to normal until after two 9-hour recovery periods, but performance on the PVT test remained low for five days. There are also influences of age on this process, in a manner that is not well understood. In one study of 64 hours of sleep loss, older subjects were more fully recovered in terms of reaction time after one night of recovery sleep than young adults, whose impairment persisted into the second night.

In terms of the sleep EEG, the first recovery night after 1–3 days of sleep deprivation is characterized by increased amounts of slow-wave sleep, sometimes with diminished N2 (stage 2) and REM sleep. Starting on the second night, as the amount of slow wave sleep declines toward normal amounts, there may be a large increase in REM sleep. This observation has been interpreted by some investigators as showing that somehow the body has a more urgent need for slow-wave sleep than for REM. This is not entirely clear, however. As mentioned earlier, recovery sleep in rats who have had very prolonged sleep deprivation is characterized by an initial profound rebound increase in REM sleep. Recovery sleep is also age-dependent, such that older people sometimes have a smaller absolute increase in slow-wave sleep and sometimes a small increase in N2 (stage 2) on the first recovery night compared to younger adults. In Chapter Five we will expand on the various changes in sleep in older people. The effects of age also highlight important topics in sleep deprivation research that continue to be explored. Among these are why some individuals are more able to tolerate sleep deprivation than others, and to what degree, if any, people are able to adjust over time to partial sleep deprivation.

SLEEP DEPRIVATION AND THE FUNCTION OF SLEEP

Sleep deprivation also provides an opportunity to consider the possible function(s) of sleep. Basically, there are several ways to look into why we sleep. One is to observe what happens when we don't sleep. Others include observing sleep in illnesses such as depression (see pages 136–9), or examining sleep from an evolutionary standpoint (see pages 47–9). One thing we can gather from sleep deprivation studies is so obvious it is easy to overlook: sleep appears to make it possible to be fully alert during waking hours. As has been touched on earlier, some viewpoints of sleep emphasize its restorative role and enhancement of other aspects of daytime function such as memory, and the role of sleep in regulation of mood, pain sensitivity, immune function, and appetite. One notion is that a function of sleep is to refine memories accumulated during the day. With new experiences, many new synapses (connecting areas between neurons) are formed. Some research suggests that, during sleep, excessive synapses are pruned back, taking away unnecessary material and making memories more precise. We also talked earlier about the glymphatic system, and the role of sleep in cleaning the nervous system of waste metabolic products. From an evolutionary standpoint, the observation that sleep lengths tend to be greatest in smaller animals suggests a role in conserving energy when there are fewer metabolic stores to draw on, or reducing exposure to predators. It seems possible that sleep may serve some (or all) of these functions, or perhaps some new one which we have not yet considered. These glimpses, without yet having a final answer, are one of the reasons the study of sleep is so fascinating.

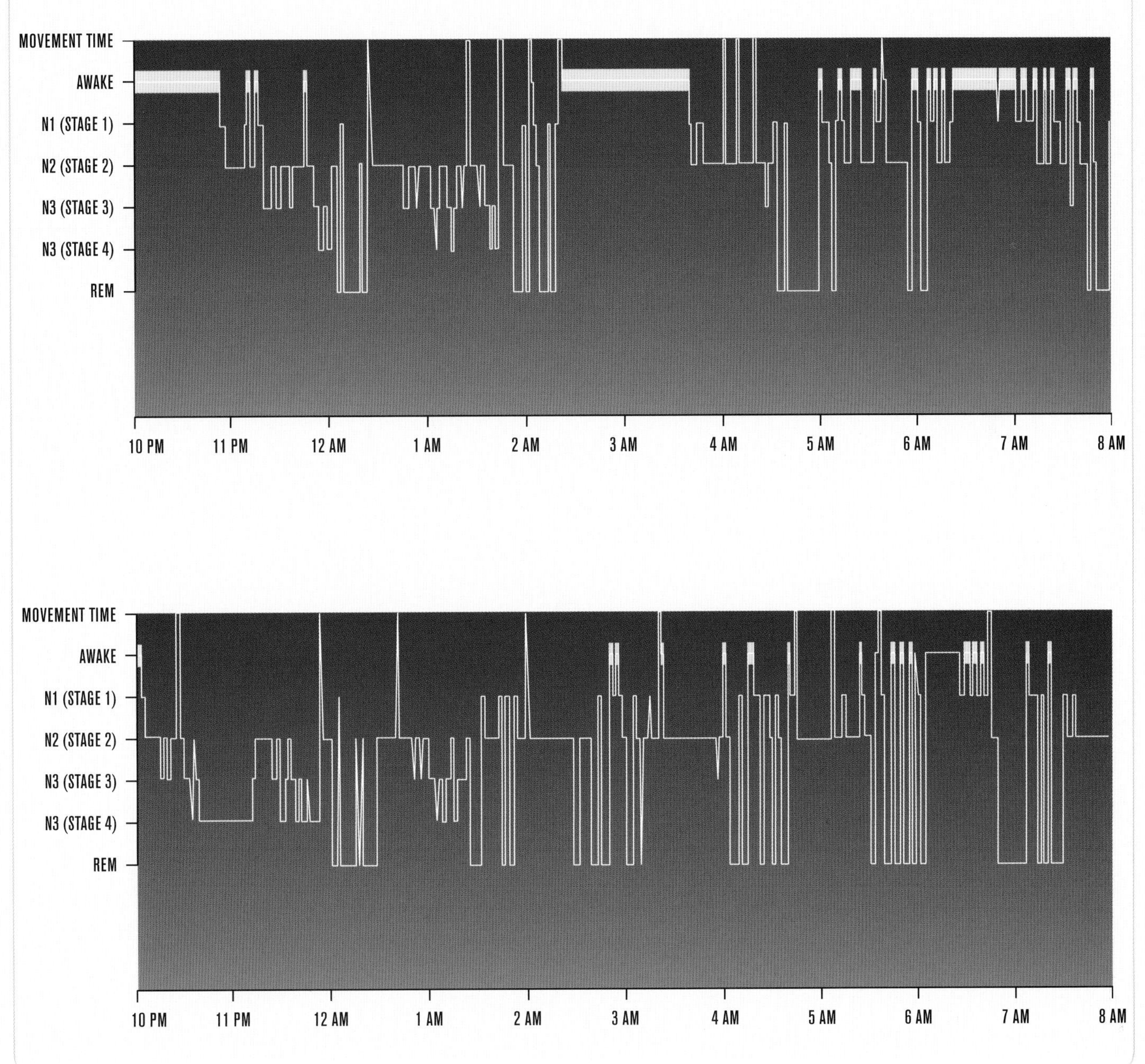

EXAMPLE OF SLEEP DEPRIVATION AND RECOVERY SLEEP
Baseline sleep (top) and first recovery night sleep (bottom) in a 65-year-old male who was sleep deprived for 38 hours. Note that on the recovery night there was a more rapid sleep onset, increased total sleep as well as stage N3 (stages 3–4 or slow-wave sleep). Adapted from Carskadon and Dement (1985).

Chapter Four

CIRCADIAN RHYTHMS AND SLEEP

One aspect of sleep, which is so obvious that we often don't think about it, is that in most species, sleep is largely confined to a certain portion of the day or night. The need to respond to alternating periods of light and dark, and the need to rest, are thought to have been important pressures in guiding the evolutionary development of species. One way of accommodating to these needs is to have a circadian clock mechanism regulating behavior and physiological processes. In this chapter we will describe how this system works, and how, if it sometimes goes awry, it may result in difficulty sleeping or problems staying awake.

HOW DOES THE CIRCADIAN SYSTEM WORK?

Description of the circadian system

"Circadian rhythms" refers to rhythms of the body clock in which one cycle has a duration of about 24 hours. It may be helpful to begin by looking at the structures in the brain that regulate the timing of circadian processes. It is thought that the master clock is contained in a small structure in the anterior hypothalamus known as the suprachiasmatic nucleus (SCN), which is described in detail in the next section. Inherent rhythmic processes can also be found in a variety of other tissues including the cornea, pituitary, liver, and lung. Although these, and indeed most tissues through the body, may have rhythmic properties, the SCN appears to provide overall regulation of them, as well as integrating circadian timing with other crucial processes including sleep and feeding. The SCN rhythm results from cycles in the activity of clock genes, which have been found in a variety of species from insects to mammals. Lesions of the SCN lead to loss of basic circadian rhythms such as those of sleep, motor activity, drinking, and cortisol secretion. When this happens sleep, for instance, appears at random times around the clock. (These rhythms can be restored by transplantation into the brain of fetal SCN tissue.) In SCN-lesioned animals, the total amount of sleep over the 24 hours remains about the same, or slightly increased because of a tendency of SCN activity to push in the direction of wakefulness.

The rhythms governed by the SCN all have a characteristic period, the amount of time from the beginning of one oscillatory cycle until the beginning of the next. In humans, the duration of the inherent period of the SCN oscillating signal has been studied by placing people into what are known as time isolation situations, usually rooms or apartments in which there are no clocks, and no clues as to whether it is day or night. This allows inherent body rhythms such as the sleep–wake cycle to become apparent. Some of the most famous studies of this type were done by the German scientists Rutger Wever (1923–2010) and Jurgen Aschoff (1913–98), who constructed a special underground bunker in Andechs, Germany, in which subjects could live for prolonged periods without any clues to the time from outside light, or changes in temperature or electromagnetic radiation. In the hundreds of studies conducted from the early 1960s to the late 1980s, in which subjects could choose their own times of light/dark and waking and sleeping, they tended to go to bed later and later, and ultimately were found to have basic body rhythms with a period of approximately 25 hours. (Later research has noted the extreme sensitivity of the SCN to even very small amounts of light, and indicates that the inherent rhythm may be closer to 24.18 hours.) If subjects stayed in this situation long

A MEMORY FOR TIME

One of the most vivid observations of the timekeeping capacity of organisms comes from Auguste Forel, a Swiss psychiatrist and investigator of insect behavior (1848–1931). In 1910, while having breakfast on the terrace of his vacation home in the Alps, he was troubled by bees that visited to sample the marmalade. They began appearing every day just before breakfast time. Ultimately driven indoors for his meal, he observed that for some days afterward the bees continued to arrive at the outdoor table at that time, and only at that time. From this simple observation he concluded that bees have a *zeitgedachtnis* (a memory for time). This is what we would refer to now as the body clock, the heart of the circadian system.

FEATURES OF AN OSCILLATING RHYTHM

Circadian rhythms (rhythms that have a duration of roughly 24 hours) are examples of oscillating systems (processes in which a measure repetitively varies around a central value across time). Rhythms involving a shorter period of time, such as that of respiration or pulse, are known as ultradian rhythms. An example of oscillating systems involving physical movement would be the motions of a pendulum from side to side around a central vertical point. There are several terms that describe this regular movement, whether it be a physical movement, or physiological processes such as a temperature rhythm or EEG waves. The cycle has an amplitude or height (the difference between its highest and lowest points); it has a period (the duration from the beginning of one cycle to the beginning of the next), and a phase (the time that the cycle begins). In biological systems, an extraneous influence (such as the effect of light on the SCN) may advance or delay the timing of a rhythm's phase (which may be expressed as degrees). The effects of such a stimulus may vary in strength or direction when given at different times of the day or night, and this is sometimes plotted in what is known as a phase–response curve.

Oscillating rhythm

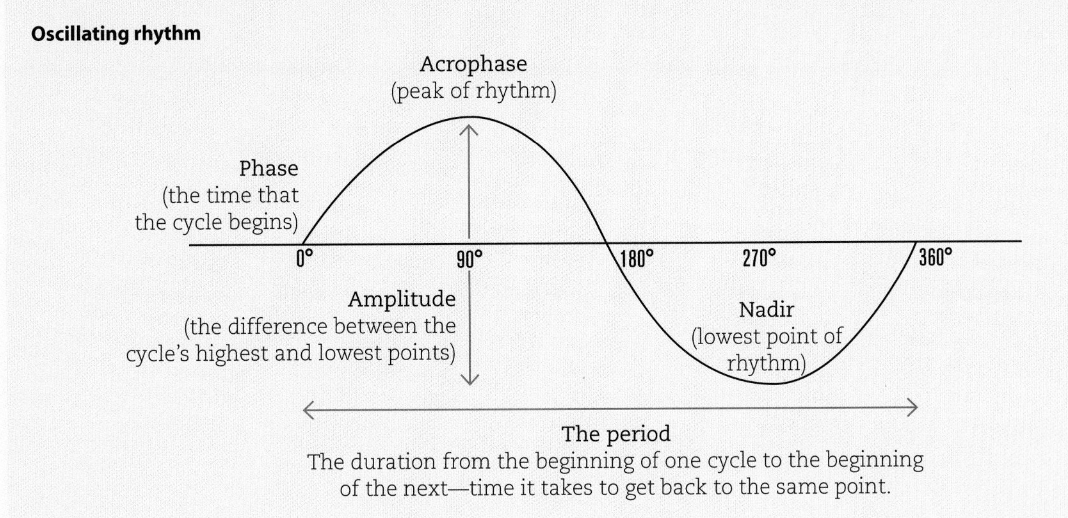

Circadian rhythm

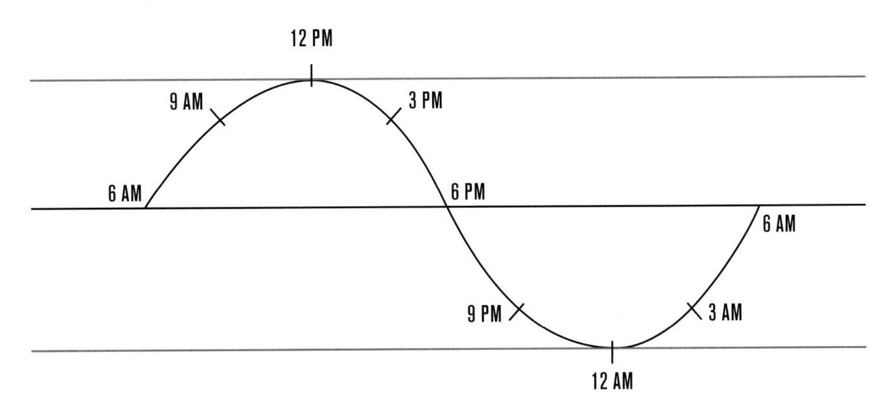

TWENTY-FOUR HOUR CYCLE

Oscillating rhythms can have a wide range of periods (durations), for instance from fractions of a second to years and much longer. When applied to a 24 -hour day, an oscillating rhythm is known as a circadian rhythm. This drawing is a sample of how the hours across the 24-hour day might appear.

enough some developed very short sleep cycles of even 12 hours, while others developed very long cycles of 60 or more hours. While this occurred, the temperature rhythm remained at about 25 hours, resulting in a situation described as "internal desynchronization," in which two rhythms (in this case temperature and sleep) no longer have their typical relationship.

Since the inherent rhythm of the SCN is in the area of 24.18 hours, and the world operates on a 24.0-hour cycle, a mechanism is in place to help re-set the internal clock mechanism every day. This is done by *zeitgebers* (German for "time-givers") in the environment, the most important of which is light and darkness. There are many other time-givers as well, including times of feeding, ambient temperature, and social interaction. The next section describes how the SCN works, but prior to that it would be helpful to look some of the terms used in describing circadian rhythmic processes, including those regulated by the SCN (see box opposite).

Neuroanatomy of the circadian system

As its name indicates, the SCN is located just above the optic chiasm, a point at which the nerve pathways, providing visual information from the retina, converge before going on to various parts of the brain. The SCN has an inherent rhythm of its own, but as we mentioned earlier is influenced by information about light, as well as by non-photic influences. The photic influences originate in a type of retinal cells called ganglion cells, using a pigment known as melanopsin, which is most sensitive to blue light. The ganglion cells can work even in the absence of functioning rod and cone cells, but in a normally operating organism the signal going to clock mechanisms is a composite of information from all three types of cells. Information about light and darkness leaves ganglion cells, and influences the SCN by two photic pathways, one direct (the retinohypothalamic tract) and one which first passes through other structures (the retinogeniculate tract).

The SCN in turn sends signals that influence sleep and behavior. One important pathway from the SCN travels by an indirect route to innervate the pineal gland, a small pine cone-shaped structure (hence its name) in a groove where the two sides of the thalamus join together. This signal, which is mediated by the neurotransmitter norepinephrine (noradrenaline), is sent when two conditions are met:

1 It is dark outside (photic information from the retina to the SCN provides this) and
2 It is night-time (as determined by the inherent clock-like mechanism of the SCN).

The hormone melatonin is then released from the pineal gland and circulates in the bloodstream, in effect providing this information (that it is dark and night-time) to the rest of the body. It also provides a kind of negative feedback to the SCN itself, which is very sensitive to melatonin.

HOW THE SCN AFFECTS OTHER PARTS OF THE BRAIN

The SCN sends a circadian signal to a number of areas. These include a pathway to the paraventricular nucleus of the hypothalamus (which in turn has an important role in appetite and anterior pituitary regulation of endocrine processes) and on to the cortex, where it promotes arousal. Another pathway goes directly and indirectly to the VLPO, which is involved in the switching between waking and sleep. Another pathway goes to the preoptic area of the hypothalamus, important in temperature regulation, sleep, thirst, and sexual behavior. The cumulative effect of these and other pathways is to orchestrate a wide range of physiologic activities. A descending pathway goes to the superior cervical ganglion (part of the sympathetic nerve system, located near the second cervical vertebra), and on to the pineal gland, regulating the release of melatonin.

Abbreviations
SCN suprachiasmatic nucleus
VLPO ventrolateral preoptic area of the hypothalamus
PVH paraventricular nucleus of the hypothalamus
SCG superior cervical ganglion
POA preoptic area
PH posterior hypothalamus

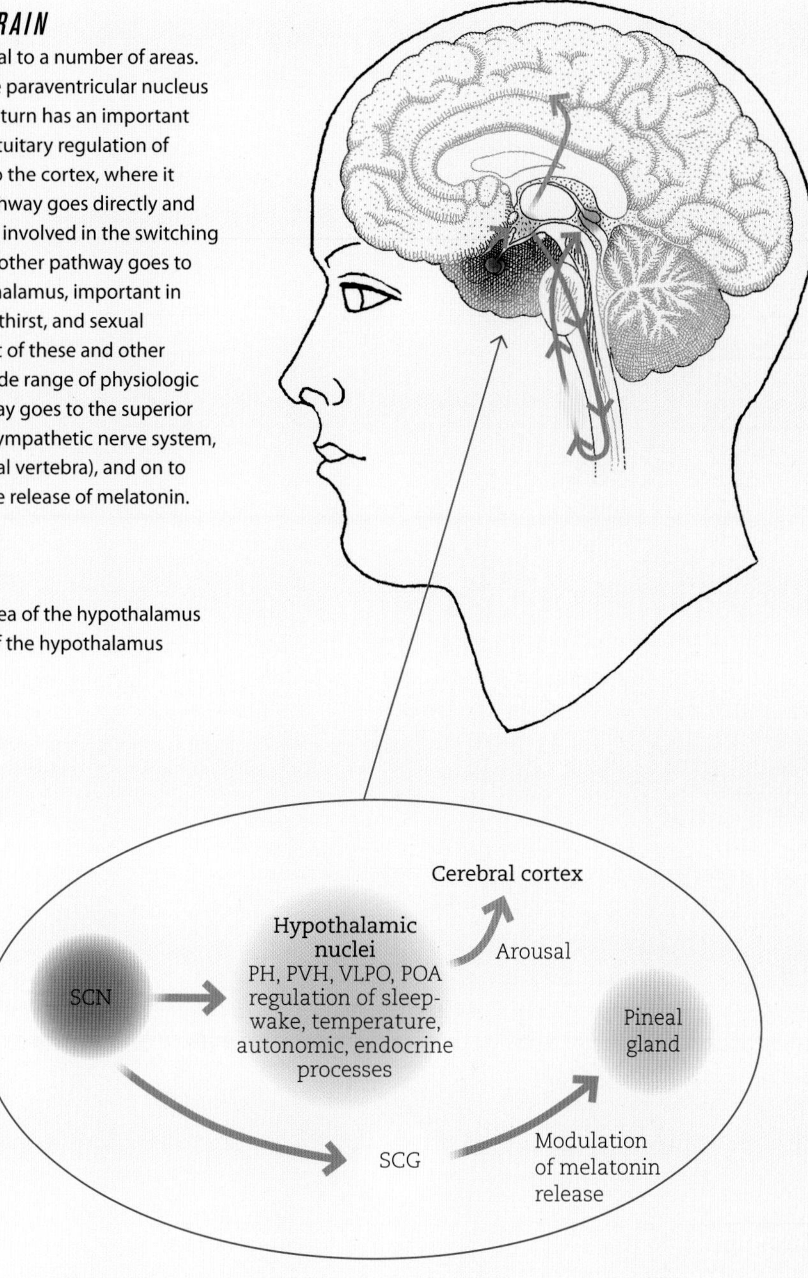

WHAT IS THE RELATION OF MELATONIN TO SLEEP?

How melatonin secretion affects sleep

Melatonin was discovered in the 1950s by doctors who were interested in the possibility that a substance from the pineal gland of cows might be useful in treating skin diseases. In the 1970s it was recognized that its synthesis in the pineal gland follows a circadian rhythm. Melatonin is chemically derived from serotonin. In mammals it is manufactured in the pineal gland by special cells known as pinealocytes, as we described earlier in response to stimulation from a pathway originating in the SCN, ultimately via the superior cervical ganglion, which releases norepinephrine. After entering the blood circulation, melatonin influences not only the SCN, but also receptors found throughout the brain and a variety of tissues throughout the body, including the liver, GI tract, arteries, heart, and kidneys. There are two major types of melatonin receptors, known as MT1 and MT2. The MT1 receptors in the SCN, when stimulated by the presence of melatonin, tend to decrease the arousing properties of the SCN. Some scientists have speculated that the melatonin-induced decrease in firing rate of some SCN neurons, as well as its effect of decreasing core temperature, produce a permissive condition conducive to sleep. MT2 receptors influence phase shifting of circadian rhythms (moving the rhythms earlier or later). These receptors have greatest sensitivity around the time of the transition from light to darkness, and hence administration of melatonin has its greatest effects when given around that time. (There is also an MT3 receptor, the function of which is not clear, but may be related to possible antioxidant properties of melatonin.) Melatonin of course affects many other body processes. One important site of MT1 receptors is the anterior lobe of the pituitary gland, which plays an important role in producing both some circadian and seasonal endocrine effects of melatonin.

Pattern of secretion

The pattern of melatonin secretion is that it is released beginning with darkness and night-time, peaks in the middle of the night, and declines toward morning. This pattern holds true regardless of whether the organism is diurnal (having the rest/sleep period at night) or nocturnal (awake and active at night). One way to think of it is that melatonin is a signal to the rest of the body that it is night-time, and that it is appropriate to engage in that particular species' habitual night-time behavior, regardless of what that is (sleep or activity). Since nights are longer in the winter, melatonin also provides information as to what season it is. This is important because it also has endocrine effects, which decrease fertility. Thus for some kinds of mammals such as sheep, goats, or horses, fertility decreases in the winter (when there is maximal melatonin secretion), and increases when nights (and melatonin secretion) become shorter.

Melatonin's feedback to receptors in the SCN provides a mechanism for altering the phase of sleep and waking. As mentioned earlier, it can directly affect the SCN, but in addition can alter its photic input due to actions on the retina and the pathway from the retina to the SCN. When melatonin is given to humans in the early evening, there is a phase advance (the rhythm shifts to start earlier) and when given in the early morning it may result in a phase delay (shifts to occur later). As we will discuss on page 162, one clinically used sleeping pill acts by stimulating melatonin receptors. There

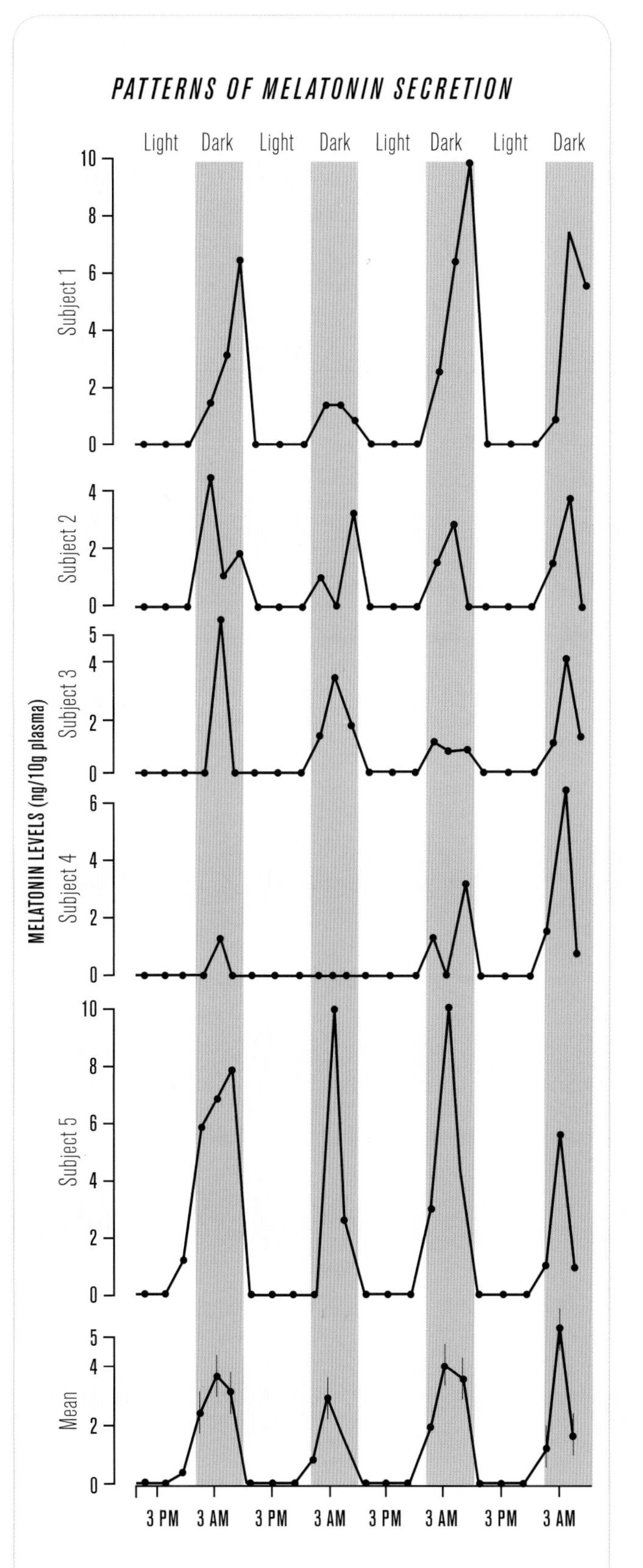

is a lack of agreement among experts on the use of melatonin itself in the types of insomnia which are not due to circadian rhythm disorders, described in Chapter Eight. One review of the scientific literature concluded that the degree of its benefits in insomnia was "unclear." Another review found that on average it decreased time to fall asleep by 7.06 minutes and increased total sleep by 8.25 minutes. It concluded that benefits were smaller than those of prescription sleeping pills, but that it might be used in view of its low cost, minimal evidence of habituation, and other factors. Melatonin also may have unwanted effects; in a "real world" survey of actual use, the American publication *Consumer Reports* found that 20 percent of people who took it complained of grogginess the next day. Dizziness and headaches can occur, but are relatively uncommon. Melatonin can also interact with various medicines including some birth control pills, anticoagulants (blood-thinning medicine), and diabetes medicines. A better established use for melatonin is in sleep disorders related to the circadian system such as delayed sleep phase syndrome and free running disorder (see "When the clock goes awry," pages 82–3).

MELATONIN LEVELS DURING THE NIGHT
Serum melatonin levels in five young adults who were kept for four days in an environment in which lights were turned off at 10:00 PM and on at 8:00 AM. The nocturnal rise of melatonin, as well as its peak at the mid-point of the period of darkness, can be seen. Adapted from Vaughn et al, 1976.

MODELS OF CIRCADIAN REGULATION

How scientists picture the workings of the sleep–wake cycle

We have just described in detail the components of the circadian system, both in terms of an endogenous pacemaker and re-setting of the pacemaker by zeitgebers such as light and signaling mechanisms including melatonin. Scientists have made various attempts to describe how this might influence the timing of sleep and waking. Perhaps the most influential model that has been developed to do this is the "two process model." It suggests that there are two major influences: the homeostatic process (which we have discussed in Chapter Three in dealing with sleep deprivation) and the circadian principle. In the model, sleep occurs at those times at which there is a coincidence of the circadian phase most permissive for sleep with a strong homeostatic drive for sleep. After sleep onset, the homeostatic pressure is dissipated, and the circadian time moves to a part which is less permissive of sleep, and then one wakes up. This model has been expanded to consider degrees of alertness and cognitive function around the clock, postulating a third interactive process known as "W," making a "three process model." Other scientists, noting that it takes a prolonged time for performance to return to normal following sleep deprivation (see page 70), have suggested that there is a second homeostatic process with a longer period. Additional models have been developed, including the "opponent process model," which suggests that the circadian pacemaker pushes toward wakefulness, which persists throughout the day until the homeostatic sleep drive is expressed at night. Another model emphasizes the average properties of neurons in the mutually inhibitory cell populations of the VLPO and monoaminergic centers as the drivers of changing between waking and sleep. These are all different ways of accounting for several basic features of the system, including a resettable pacemaker, input of time-giving stimuli, and a homeostatic drive for sleep, which ultimately produce the observed circadian rhythm of sleep and waking.

Genetic regulation of circadian rhythms

The first gene recognized to have a role in regulating circadian rhythms, known as PER, was described in fruit flies (drosophila) in the early 1990s. Various mutations of PER resulted in the flies living on short (18–20 hour) or very long (28–30 hour) daily rhythms. By now at least seven circadian-related genes have been identified in drosophila, and one strain has been bred which has great difficulty sleeping and has been proposed as a model for human insomnia. In mammals, attention has been focused on the rhythmic activity of three period genes (PER1, 2, and 3) and two genes of a type known as cryptochromes (CRY 1 and 2). The associated proteins PER and CRY provide rhythmic negative feedback on the activity of their parent genes. Variations in the area of PER1 have been related to changes in up to an hour in the timing of the peak (acrophase) of the human daily motor activity rhythm. In principle, such information might be used in a variety of ways, including individualizing scheduling of shift work, and could potentially bear on timing of medical treatments. As mentioned in Chapter Three, different forms of PER3 have been related in humans to varying degrees of subjective sleepiness and memory function after sleep deprivation.

HOW CIRCADIAN AND HOMEOSTATIC PROCESSES INTERACT

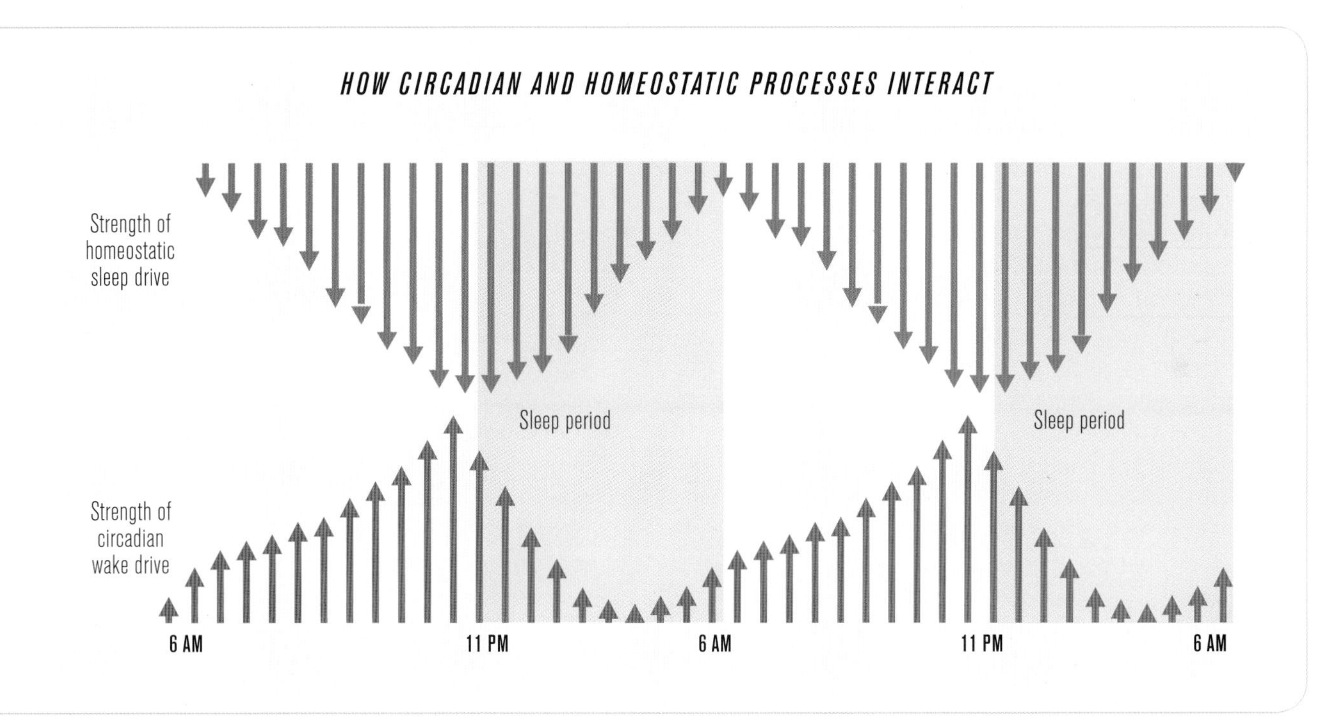

CIRCADIAN AND HOMEOSTATIC DRIVES
In the two-process model of sleep regulation, the timing of sleep onset depends on the interaction of the homeostatic drive (which leads to a tendency toward sleep the longer one is awake) and the circadian system. As seen here, homeostatic influence increases from the time a person awakes in the morning. Sleep onset occurs around the time that homeostatic drive is maximal, and the circadian system is in a phase which is more permissive of sleep. Once a person is asleep, the homeostatic drive begins to dissipate. Waking onset occurs when homeostatic drive is low, and the circadian system is less permissive of sleep. A modification of this view, the opponent process model, suggests that the circadian system not only provides timing, but also pushes in the direction of wakefulness.

AN UNUSUAL ZEITGEBER

Earlier in this chapter we introduced the concept of a zeitgeber—a time-giver such as sunlight, which helps re-set the circadian clock every day. During his time serving as ambassador to France, American revolutionary figure, diplomat, and scientist Benjamin Franklin (1705–90) proposed an unusual zeitgeber. To maximize time awake when it was light outside, and to conserve candle wax at night, he unsuccessfully proposed to the government that they fire cannons at dawn in Paris to awaken the citizenry.

WHEN THE CLOCK GOES AWRY

Sleep disturbances related to circadian system function

The circadian system is complex, and depends on the interaction of many different parts of the nervous system. It is not surprising, then, that sometimes it does not function optimally, either in normal conditions or when met with challenges such as changing sleep times. These situations may manifest as clinical sleep disorders.

Jet lag

Jet lag and shift work are similar in the sense that both result from a mismatch between a person's internal rhythm of sleep/waking and the external expectations of the environment. Thus a person flying eastward would be expected at the new locale to go to bed before his or her circadian processes are ready for sleep. The result is that (s)he would tend to stay up later than the expected bedtime, and get up later in the morning, than the traditional hours in the new environment. Conversely, a person flying westward would tend to go to sleep earlier, and then get up earlier, than is customary locally. Jet lag is further complicated by sleep deprivation related to traveling, and sometimes by alcohol consumption en route. Thus it can result in sleepiness and impairment of skills such as driving at the new-destination daytime, and insomnia when trying to sleep at the new bedtime. Gastrointestinal symptoms are common as well. In general it takes about one day for each time zone crossed to adapt, though for some unhappy travelers it can go on for substantially longer times. Westbound travel is usually a little easier as the adjustment needed to the new environment is a phase delay, which fits naturally with the tendency of the basic SCN rhythm to be a little longer than 24 hours; eastward travel requires a phase advance, which is often more difficult. The one piece of good news is that in a sense jet lag is a little easier to deal with than shift work, because in jet lag the light and environmental expectations in the new environment are usually of more help in synchronizing.

TRAVEL ACROSS TIME ZONES

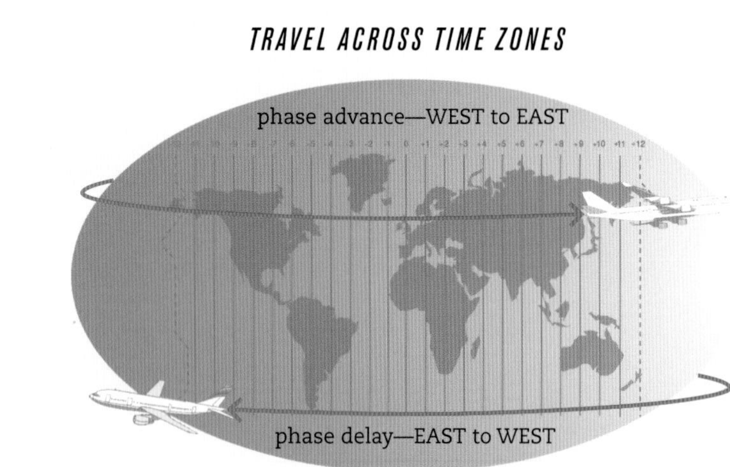

DEALING WITH JET LAG

When traveling westward, one becomes phase advanced relative to the new environment, and needs to have a phase delay to adjust to the new situation. The opposite is true for eastward travel. Using the model of traveling between East coast United States to Europe, crossing six to seven time zones, one can begin by trying to partially adapt before the trip. When going eastward, for two or three nights before the trip try to go to bed and get up earlier, while reducing exposure to light in the evening, and increasing exposure to bright light in the morning. Conversely for a few nights before going westward, try staying up later, while increasing bright light in the evening and reducing exposure to light in the morning.

SHIFT WORK
About 16 percent of American workers have shifts other than regular daytimes. Many people can tolerate these schedules without overt illness, but some can develop shift work disorder, which in addition to sleepiness and sleep disturbance can include a higher risk of accidents, mood disorders, headaches, fatigue, gastrointestinal problems, and other difficulties.

As of this time, there are no medicines specifically approved by the US FDA for jet lag. Traditional sedative/hypnotic medications such as the GABA agonists (see pages 160–3) may help improve sleep at the new bedtime. Generally they are thought to not aid in adjustment to the new time zone, though one study of the sleeping pill triazolam reported some benefit in a simulation of a westward flight across eight time zones. Melatonin has been used in jet lag, and some very limited data suggest that the melatonin agonist ramelteon (see page 162) may aid going to sleep after eastbound travel. There are data to suggest that armodafinil (see pages 128–9) may improve wakefulness after eastward travel through six time zones. (Again, none of these drugs have specific regulatory approval for use in jet lag.) Caffeine can be of some help in promoting wakefulness during the daytime. During jet lag, one is particularly sensitive to the sedative and other effects of alcohol, which should be avoided or used with caution. And the very real propensity toward auto accidents while having jet lag needs to be kept in mind.

Shift work

About 84 percent of American workers are in daytime-only jobs; the remaining 16 percent work with a variety of schedules including night and rotating shifts. A significant minority (14–32 percent of night shift workers and 8–26 percent of rotating shift workers) can develop what is known as shift work disorder (SWD). Symptoms can include significant insomnia and sleepiness, even though the person involved may have a predictable sleep schedule, adequate available time in bed, and no other sleep disorders. It is not clear why some people can adapt less easily to shift work. As mentioned in Chapter Three, some forms of the gene PER3 can lead to increased difficulties with sleep loss. It has also been found that those who are more "morning people," as well as older people, can less easily adapt. In addition to the insomnia and sleepiness, SWD takes a number of tolls on one's health, including higher rates of depression, ulcers, and accidents. Shift work in general may also be associated with higher rates of heart disease.

Several scheduling approaches can make it a little easier to deal with shift work. These include staying on particular shifts for longer periods of time (measured in weeks) rather than changing rapidly, and changing shifts in a forward direction (day to evening to night), following the natural tendency of the inherent SCN rhythm, which is longer than 24 hours. The use of bright lights in the evening to induce phase delays or in the morning to induce phase advances can help, but are difficult to do in real world settings. Melatonin has been used as well, but its effectiveness is limited in the absence of good control of lighting. A nap followed by caffeine before reporting for night work has been reported to be of some help in alertness.

The use of traditional stimulants such as dextroamphetamine or methylphenidate is to be discouraged due to their significant side effect profiles and risk of dependence. Modafinil has been reported to improve wakefulness and performance in shift work settings, and its longer-acting isomer armodafinil has been approved by the US FDA for use in shift work disorder.

Delayed sleep phase syndrome

Individuals with this difficulty have a chronic (more than three to six months) stable, but delayed, sleep–wake pattern relative to that of the community. Typically, they have trouble falling asleep until after 2:00–3:00 AM, and often much later, and if given the choice will sleep until the late morning. Once they do fall asleep, it is basically normal, and if allowed to sleep until late morning, they will then feel fully alert. The difficulty, of course, is that such a schedule does not fit well with society's plans for going to work or school. The result of having to get up early to meet these goals is that they are sleepy, often irritable, and feel poorly in general. Thus delayed sleep phase may result in complaints of insomnia (difficulty going off to sleep) or morning sleepiness. Though delayed sleep phase syndrome can occur at any age, it often occurs in adolescence, where it may result from a combination of habitual behavioral patterns and the delay in the sleep rhythm characteristic of that stage of development (see pages 92–3).

Possible causes of delayed sleep phase syndrome include a decreased effectiveness of the normal entraining process by light, and possible defects in homeostatic regulation as well. A familial form has been identified which is inherited as an autosomal dominant trait, and it may be associated with a variance in the human Per3 and Clock genes. Treatment is often by making small phase advances (going to bed a half hour earlier at a time). In severe cases in which a person can't sleep before perhaps 6:00 AM, one can do chronotherapy, in which (s)he is moved around the clock in a forward direction, going to bed 3 hours later each day until coming around to a more conventional bedtime. Melatonin has also been used in the treatment for delayed sleep phase syndrome.

Advanced sleep phase syndrome

In this common condition, people chronically tend to go to sleep several hours earlier, and wake up several hours earlier, than in the general population. The presenting symptoms are sleepiness in the evening,

PHASE SHIFT BETWEEN TWO WAVES

In phase delay syndrome, the sleep period is in a stable but delayed rhythm compared to most of the population. In terms of oscillating rhythms (see page 75), a phase-delayed rhythm (dashed line) is seen to the right of the basic rhythm (solid line). The amount of the phase delay is often expressed in degrees, in this case 30 degrees, which, if the period (duration) of the full wave is 24 hours, would be about 2 hours.

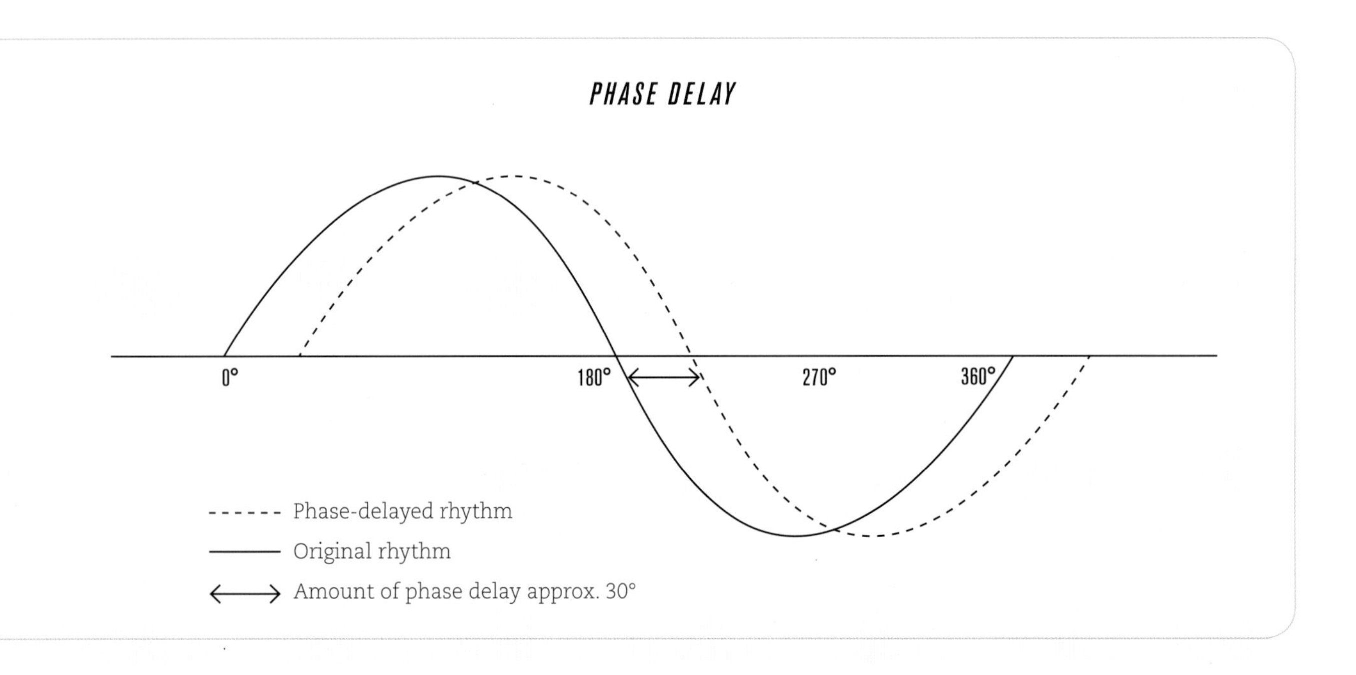

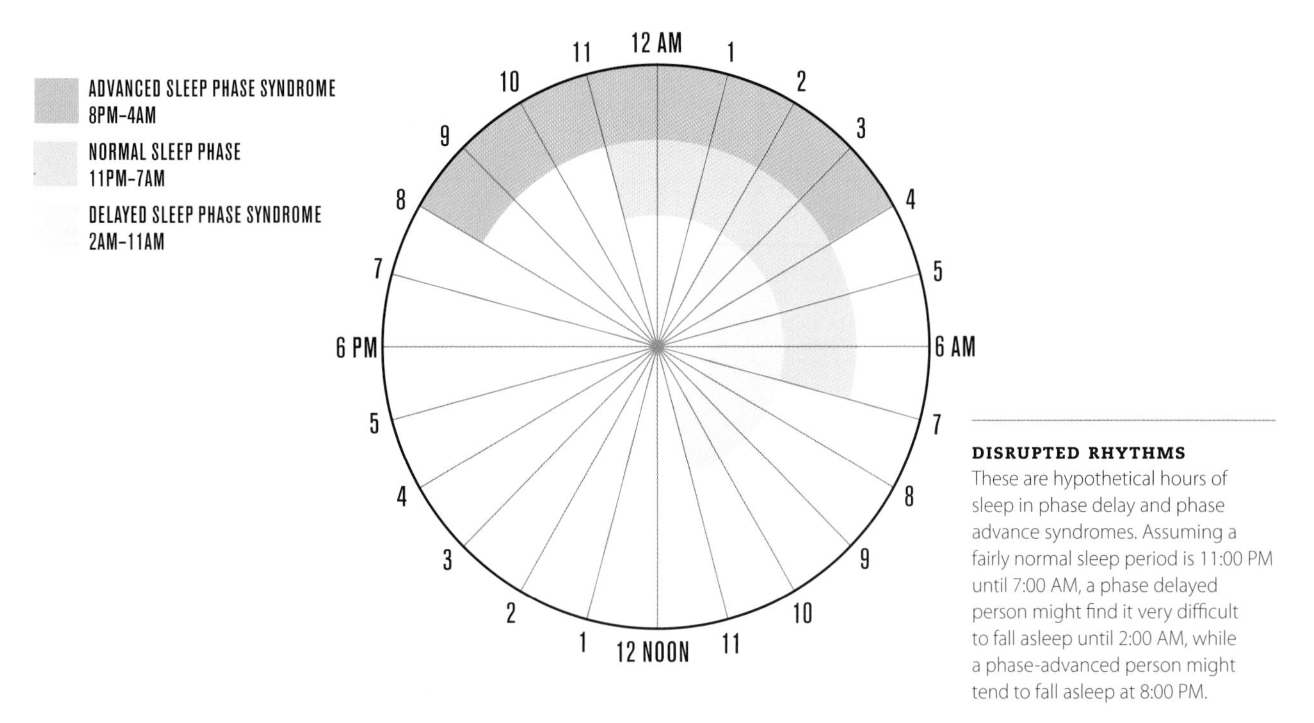

DISRUPTED RHYTHMS
These are hypothetical hours of sleep in phase delay and phase advance syndromes. Assuming a fairly normal sleep period is 11:00 PM until 7:00 AM, a phase delayed person might find it very difficult to fall asleep until 2:00 AM, while a phase-advanced person might tend to fall asleep at 8:00 PM.

and early morning awakening. It often is associated with middle or older age, and may be due to heightened sensitivity to the phase advancing properties of morning light, or to a shorter endogenous circadian rhythm. Treatment is usually by exposing the person to bright light in the early evening.

Irregular sleep–wake rhythm

People with this disorder sleep at multiple irregular times around the clock, though their cumulative 24-hour sleep time may be relatively normal. This is often found in elderly patients with dementia, as well as after brain injury and in institutionalized children. It may be due to difficulties in the circadian regulatory system, often in combination with diminished social and light cues in institutional settings. This syndrome should be distinguished from the pattern in people who voluntarily choose to live a lifestyle of very irregular sleep habits. Treatment is by maximizing social and light cues, using bright light during the daytime and structuring the waking hours with social events and exercise.

Non 24-hour sleep–wake disorder (free running disorder)

Free running disorder is almost exclusively seen in blind people, though rare cases in the sighted have been reported. These individuals have an endogenous rhythm of more than 24 hours, which is not re-set each day by time cues such as light. As a result, their internal schedule is continually moving forward around the clock. They typically complain of insomnia and daytime sleepiness, which get worse every few weeks, when their rhythms are most out of sync with society. Some doctors recommend melatonin around 9:00 PM on a regular basis. The medication tasimelteon is approved for this disorder in blind people in the United States and European Union.

In summary, then, the circadian mechanisms involve a highly orchestrated system. Sometimes the organization of this system or its integration with the outside world are faulty, resulting in problems of insomnia or sleepiness.

Chapter Five

A LIFETIME OF SLEEP

So far we have touched upon a number of influences on the amount and timing of sleep, such as circadian and homeostatic processes. These mechanisms act on a person who already has built-in predispositions for certain amounts of sleep due to his/her age. This chapter describes sleep in the newborn, showing how in the first few months the recognizable patterns of sleep begin to form, how they develop across childhood and adolescence, and progress across adulthood and older age. It will also be seen that circadian rhythms of sleep and waking alter across the lifetime, for instance becoming later in adolescence and earlier in old age. We will also lay the groundwork for seeing how some illnesses of sleep are characteristic of different ages, as presented in Chapter Seven.

WHAT IS SLEEP LIKE IN THE FIRST YEARS OF LIFE?

Sleep in infants and children

Sleep in infants and children can be seen as the beginning of a process that changes across the lifetime. In the next sections we will first describe some overall patterns, and then look at each major age group in turn.

General trends

A few generalizations can make it easier to picture how sleep evolves. Total sleep time is highest in infants and children, declines but then levels off in adolescence and adulthood, and then begins to decline again in older age. The percentage of sleep occupied by the individual sleep stages changes as well. REM sleep percentage follows a pattern similar to that of total sleep: very high in infants and children, declining until adolescence, remaining relatively constant during adulthood, then declining in older age, but with substantial variability. Slow-wave sleep is highest in infants and children also, then the decline slows after adolescence but continues slowly across adulthood and into old age. In contrast, the number of awakenings is relatively low early in life, and increases

DEVELOPING PATTERNS IN INFANCY
The first two years of life are a dynamic period with a variety of changes in sleep. Among these are the decline in total sleep time around the 24 hours, the consolidation of sleep primarily to the night, and a reduction in active sleep.

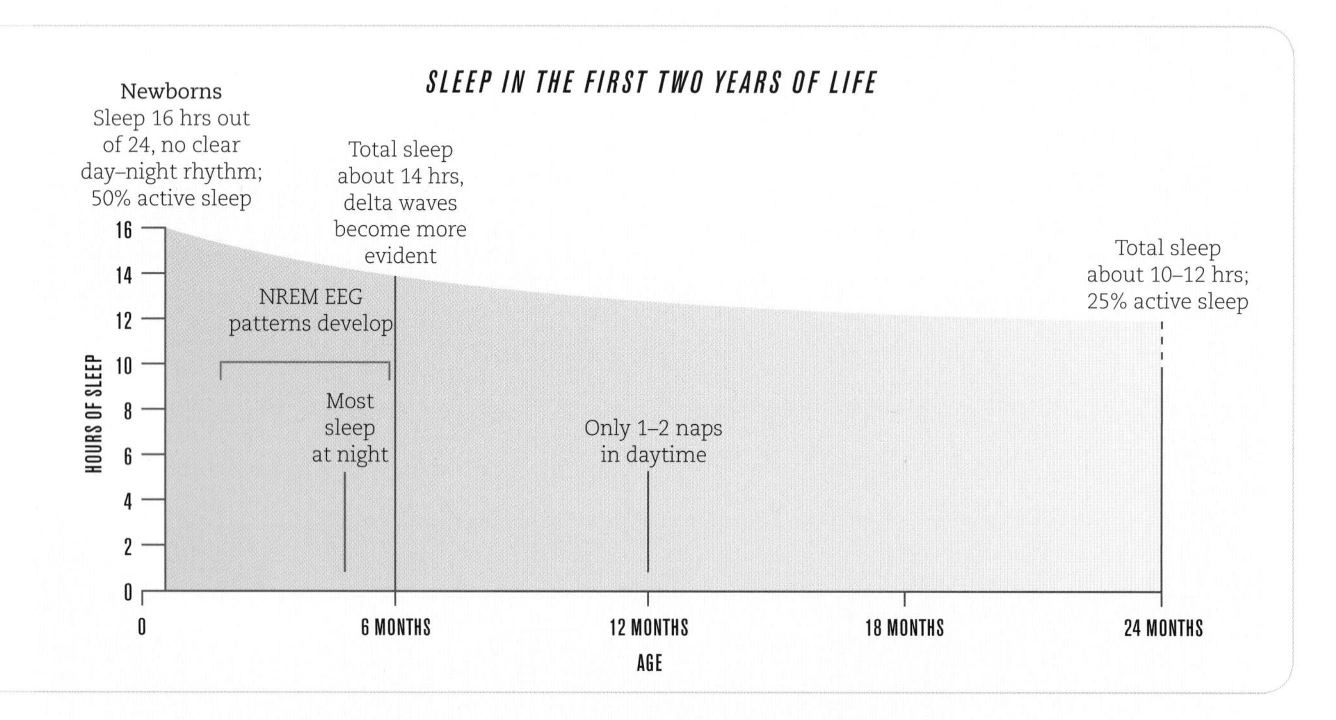

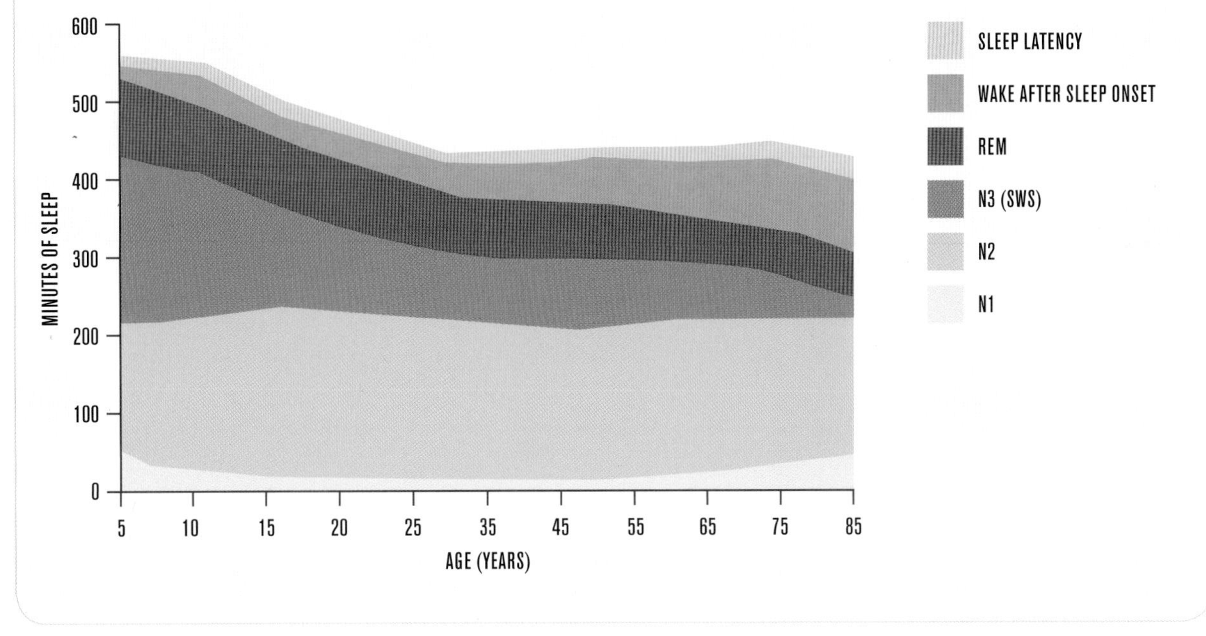

THE EFFECT OF AGE ON SLEEP
Sleep measures in healthy individuals, taken from an analysis of 65 studies with a combined group of over 3,500 people. Among the many interesting observations was that a decline in total sleep time in children and adolescents was noted only in studies performed on school days. Note the significant decrease in stage N3 across the lifetime. By age 85, sleep was characterized as having more stage N1 and wake time after sleep onset and longer sleep latency, perhaps corresponding to the greater frequency of complaints of poor sleep in older age. Adapted from Ohayon et al (2004).

gradually across the lifetime. Complaints about poor sleep rise across adulthood, from roughly 20 percent of young adults to over 50 percent in those over 70. This high rate of sleep difficulties in the elderly in part may reflect the presence of a number of clinical sleep disorders and medications.

Total sleep and rhythmicity of sleep in infants and children

The sleep of newborn infants has little day–night difference and, though the quantity is relatively great (in the area of 16 hours), is scattered around the clock. After about two months, a discernible circadian pattern begins to emerge, though initially it is free running, that is, responding to an internal rhythm and not being re-set daily by light and dark (see Chapter Four, pages 72–85). At around four to five months, sleep begins to be entrained by light and darkness, and (thankfully for parents) the major sleep period begins to consolidate at night. By one year there are usually only one or two naps during the daytime. Napping at least once during the day declines from over 95 percent at 1.5 years, to about one-third of children at four years, with primarily only night-time sleep at school age. Total amount of sleep declines to about 14 hours at six months of age, then 10–12 hours by age two, and continues to drop, to about 8 hours, in the second decade of life (in conditions in which there are no social constraints such as needing to get up for school).

Sleep stages

In terms of sleep stages, newborns tend to go directly from waking into what is known as “active sleep,” which with time develops into REM sleep. Active sleep at this point comprises perhaps 50 percent of sleep, and is even higher (up to 80 percent) in premature babies. After the first three to four months, active sleep declines, and babies go from waking into quiet sleep. Active sleep continues to decrease, to about 20–25 percent by two years of age. The duration of the REM–NREM cycle is also very short, in the area of 50–60 minutes, compared to about 90 minutes in adults. By two to four months, sleep spindles are evident, as are K-complexes at four to six months. By six to seven months delta waves become more evident. Most striking is the development of slow-wave sleep, which becomes so prominent in infants and children that they often miss what would have been the first REM period. During slow-wave sleep, children sleep very deeply, in the sense that it takes a relatively loud sound or other stimulation to awaken them. Parents may have the experience of picking up a sleeping child in one room and carrying her to the bedroom without awakening. Not evident to parents but as seen in the laboratory, alpha EEG activity (see pages 12–13) has been slowly making its appearance until it becomes well developed by about age eight.

Sleep-related endocrine processes

Not only sleep itself, but sleep-related endocrine processes develop in infancy, and ultimately change in childhood and adolescence. In Chapter Six, we will discuss sleep-related secretion of growth hormone (see pages 102–3). In newborns, daytime and night-time blood levels of growth hormone are about the same, as are amounts in active and quiet sleep. By the third month, around the time that quiet sleep comes to be more prominent, growth hormone levels in waking have declined considerably and are much lower than those in quiet sleep. In prepubertal children there is significant sleep-related secretion compared to daytime, and in adolescence there is an increase in both night-time and daytime secretion.

FLUCTUATING SLEEPING PATTERNS
As many exhausted parents will attest, in newborns there is little day–night difference in sleep. Typically by four to five months, the major sleep period begins to be primarily at night.

Childhood sleep disturbances

Some sleep disorders tend to occur at different ages, and a number of these are associated with infancy and childhood. Infants may have rhythmic body rocking or hitting their head on the side of the crib. This may be alarming but is often a benign process that is managed by padding the crib. Sleepwalking is not uncommon, and in most cases is a benign process of NREM sleep, which can appear more often at times of stress; much less commonly it may be a manifestation of nocturnal seizures, which are of medical concern. Obstructive sleep apnea (see pages 110–11) can occur in children. It may manifest in other ways than in adults; instead of obvious sleepiness, behavioral disturbances including those similar to attention deficit hyperactivity disorder, or aggressiveness, or excessive shyness may be evident. In contrast to adults, removal of tonsils is often the most helpful procedure, and positive airway pressure can also be used. These disorders, as well as nightmares and night terrors, are discussed in Chapter Seven (pages 130–3).

Another factor which is associated with poor sleep in children is the use of cell phones and tablets at bedtime. An analysis of 20 studies found an association with shorter sleep, a decline in sleep quality, and increased daytime sleepiness. This was true even when children had access to the devices but did not necessarily use them. Studies have not yet been done to see the effects of removing cell phones and tablets from the bedroom, but in the meantime, it is something to be kept in mind.

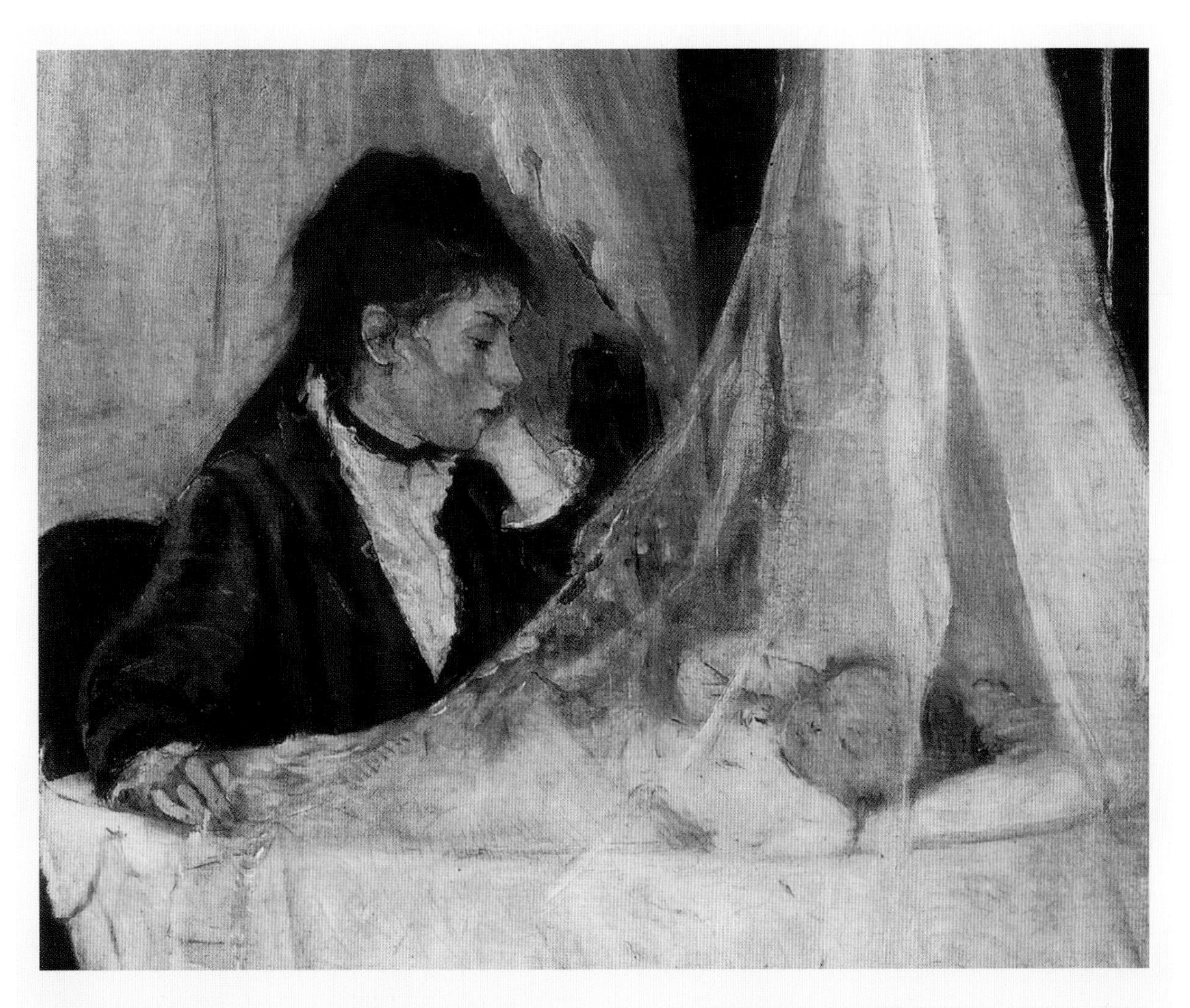

EMERGING PATTERNS
In *The Cradle*, 1872 by Berthe Morisot (1841–1895), a mother watches over her sleeping infant. Parents will notice changing sleep patterns during their child's first years of life.

FROM THE PARENTS' VIEW: SLEEP OF INFANTS AND CHILDREN

Parents will typically observe the sleep of newborns as having no rhythm or significant day–night difference. By around two months, the first elements of a rhythmic pattern emerge, but are not closely related to the environment. By four or five months, sleep becomes more influenced by light and darkness, and most sleep will occur at night. In newborns and during the first few months, parents may sometimes observe the behavioral manifestations of active sleep: periods in which there are occasional grimaces and quick twitching of the limbs, sometimes with respiration that is slightly more irregular than in quiet sleep. The total amount of sleep will be seen to be declining from up to 16 hours per day in newborns to 10–12 hours at age two, continuing to drop until adolescence. The tendency to fall asleep will occur during the mid-evening in children, in contrast to the tendency to stay up later which will appear in adolescence (see pages 92–3).

HOW DO ADOLESCENTS SLEEP?

Sleep in a changing body

It has been said that it is easy to tell when children enter adolescence (the second decade of life): instead of having difficulty getting them to stay in bed a bit later in the morning, now parents can't get them up. One aspect of this may be that higher levels of stimuli are necessary to awaken adolescents than adults. From an evolutionary point of view, it has been argued that nature finds it safe to let them sleep deeply because parents take on the duty of protection from predators. A second, perhaps more significant aspect is the natural tendency in adolescence for both the beginning and the end of sleep to shift to a later time (a "phase delay"). As far as we can tell, this is a very basic process, not necessarily related to a specific culture, as it has been observed in the United State, Europe, and Asia, and is seen in pre-industrial cultures as well. Later sleep times appear to occur before the onset of puberty (sexual maturation). Unfortunately, this shift in the timing of sleep conflicts with the necessity of getting up early for school, which often starts at 8:30 AM or earlier. As the need for sleep appears to remain the same or even increase, at least during older adolescence, the result can be chronic sleep deprivation, which potentially can have effects on alertness and the ability to retain information (see pages 56–7). On weekends, younger adolescents tend to sleep about the same amount, but older adolescents often sleep one or one and a half hours later. Since there is a tendency among older adolescents to go to bed later on weekend nights, the result for the week as a whole is an irregular sleep pattern, which may predispose to developing sleep disorders. A recent study of adolescents also found that those with later bedtimes during the weekdays were more likely to have an increase in Body Mass Index, even when compared to others with the same amounts of total sleep time and exercise. The reasons for this are not clear, though there was some association with intake of fast food. The decline in total sleep time on school nights is seen

SLEEPY TEENS
In adolescence, there is a tendency for sleep to shift to a later time. When this combines with the need to get up early to go to school, the result can be chronic sleep deprivation and daytime sleepiness.

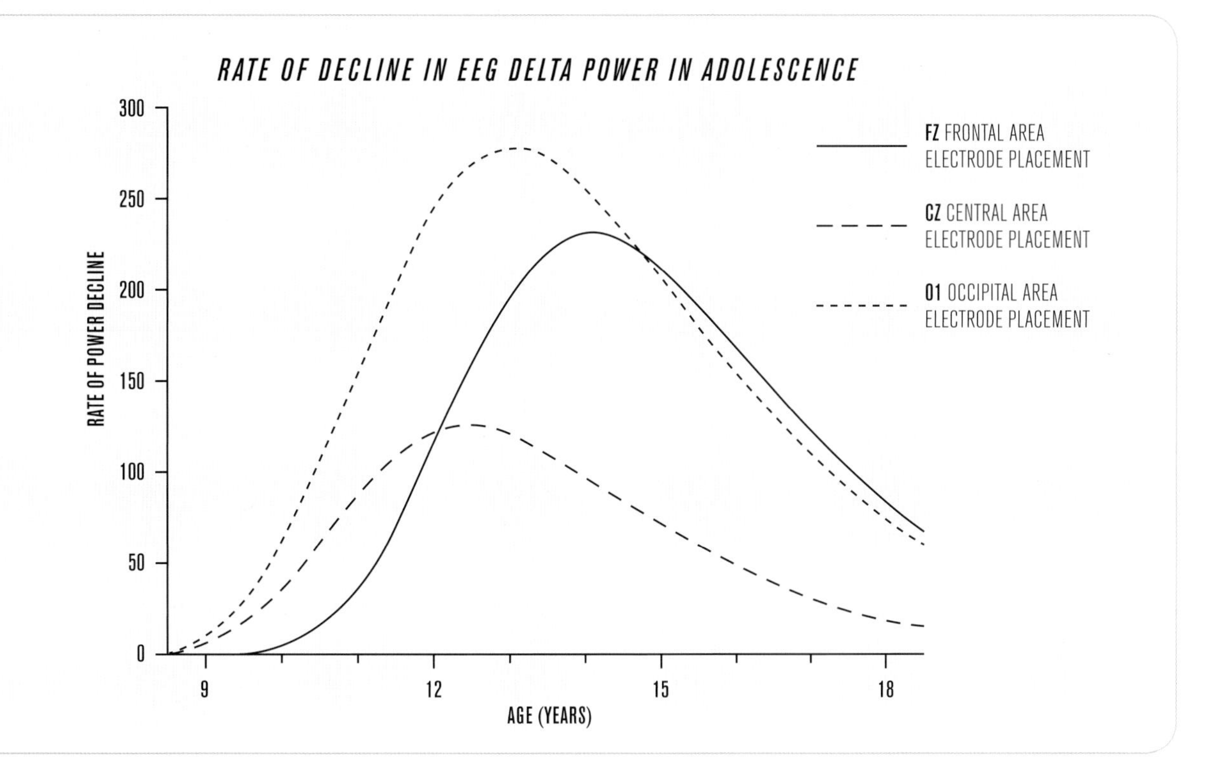

in sleep laboratory studies as well. REM sleep declines slightly, until it levels off in young adulthood. In terms of survey data, girls report that they have a longer period awake in bed before going asleep, but then sleep longer, than boys. Although this is the overall pattern, among those who do develop insomnia, there is some association with menarche (the onset of menstrual function). Boys who develop insomnia tend to start at a little earlier age. In adolescents with insomnia, electronic analysis of the EEG indicates higher amounts of beta range activity (see pages 12–13) during NREM sleep. This is seen in adults with insomnia as well and thought to reflect higher levels of cerebral cortical arousal. In Chapter Seven we will talk about the concept of insomnia involving excessive arousal around the clock, which can be seen in other ways including brain imaging studies.

Initially in adolescence, slow-wave sleep continues to be high, as in childhood. Indeed, early adolescents may miss their first REM period, possibly because of the large amount of slow-wave sleep. Later, as slow-wave sleep declines, REM begins to appear

DECLNE IN STAGE N3 (SLOW-WAVE SLEEP)
The decline in stage N3 (slow-wave sleep) during adolescence is very marked, dropping roughly in half between ages 10 and 20. This is also reflected in electronic analyses of the sleep EEG, as seen in the rate of decline of electronically-calculated power in the delta wave band. Adapted from Feinberg et al (2011).

earlier. This decline in slow-wave sleep is perhaps the most striking change in the sleep EEG during adolescence. It may drop by 50 percent between ages 10 and 20. This generally occurs earlier in girls than boys. The decline in visually scored slow-wave sleep seems to correlate well with the progressive degree of sexual maturation, though some other measures show it more closely related to age. The American sleep researcher Irwin Feinberg (1928–) has speculated that the decline in slow-wave sleep represents part of a major reorganization of the brain during adolescence, during which there is a decline in the number of interconnections between cortical neurons ("synaptic density"), decreased plasticity of the brain, and the beginning of more adult forms of thinking processes.

HOW DOES SLEEP CHANGE IN ADULTHOOD?

Sleep from young adulthood to older age

After the tumultuous changes in sleep across childhood and adolescence, adult sleep becomes much more stable. The tendency toward a phase delay in sleep onset, which had been characteristic of adolescence, is no longer present. The amount of sleep is relatively stable from mid-twenties until mid-sixties, and the typical amount is 7–8 hours, though many people sleep substantially less or more (see pages 28–31).

Young adulthood

In young adults, N2 (stage 2) comprises about 50 percent of the night, REM 25 percent, slow-wave sleep 20 percent, and N1 (stage 1) 5 percent. Typically there are four to five REM sleep periods each night, beginning with relatively short ones early at night about 90 minutes after sleep onset, and getting longer as the night progresses. There is relatively more slow-wave sleep at the beginning of sleep, declining as the night goes on. Depth of sleep (in terms of how loud an auditory stimulus it takes to awaken a person) remains greatest in slow-wave sleep, but the absolute amount of stimulus needed for awakening is less than that in childhood and adolescence. The number of awakenings each night is relatively small. The majority of young adults feel that their sleep is restful, and complaints of insomnia are certainly present (about 10 percent in a Canadian study in young adults up to age 34) but relatively lower compared to later in life. For many young adults, though, an important issue can be sleep deprivation due to lifestyle, which can result in daytime sleepiness. Stress and anxiety can play a relatively greater role in sleep disturbance of young adults. Some of the major illnesses that result in disturbed sleep, such as periodic limb

SLEEP AND THE MENSTRUAL CYCLE

The menstrual period may also affect sleep in women. About half describe awakening earlier and feeling less refreshed, and taking longer to fall asleep, just before and during menstruation. Among women who have premenstrual syndrome (which can additionally include bloating, headaches, mood changes, cramps, and other symptoms) either insomnia or excessive sleepiness may also occur. Some doctors use selective serotonin reuptake inhibitor medications to ease symptoms, though clinically, benefits on the sleep aspects seem modest. Cognitive behavioral therapy has been reported to be of benefit. Some authors have recommended calcium supplements or the amino acid L-tryptophan.

Menopause may also be a time of sleep disturbance. Hot flashes (a feeling of heat in the face and upper torso) are often associated with awakenings. These may be relatively brief (often less than half a minute) but cumulatively they lead to a sense of restless sleep. In women who are aware of hot flashes, there may be as many as 15–20 per night. In general women with symptomatic hot flashes have decreased sleep efficiency, and more time awake intermittently during the night. The use of hormone replacement therapy to ameliorate sleep disturbance and other symptoms is a complex issue, which must balance possible benefits against potential medical complications. (One major study, the Women's Health Initiative, suggested an association with breast cancer, heart disease, and vascular dementia.) For this reason, many physicians are reluctant to recommend hormone replacement therapy, or suggest that women use it for only brief periods of time. Some doctors use selective serotonin reuptake inhibitor medications such as paroxetine. Non-drug therapies such as relaxation therapy and stress reduction have been employed, together with lighter bed coverings.

movement disorder and obstructive sleep apnea have relatively lower incidence, which will later rise in middle and older age. A significant exception to this is narcolepsy, which causes both excessive sleepiness and disturbed nocturnal sleep, and which typically begins in adolescence and young adulthood (see pages 124–5).

Middle age

As adulthood progresses up to age 65 or so, several changes begin to appear. Total sleep remains relatively stable, but the number of awakenings progressively increases. Though the sharp decline in slow-wave sleep seen in adolescence gradually abates, a slower decline continues across adulthood. During this period, amounts of REM sleep remain relatively steady. Clinically, the rate of complaints of disturbed sleep gradually rises, reflecting at least in part the increase in awakenings. As mentioned earlier, sleep disorders such as periodic limb movement disorder and obstructive sleep apnea become more frequent as well. Medical illnesses and the use of medications which can disturb sleep are also on the rise. Beta blockers for high blood pressure or cardiac arrhythmias, and some cholesterol-lowering statin drugs, for instance, may disturb sleep. Another example would be the group of drugs known as "ACE inhibitors" which are used for high blood pressure, but which in up to one-third of patients can produce a dry cough around the clock that can disturb sleep. Many adults, due to lifestyle, continue to get less sleep than the roughly 7.5–8.5 hours which are generally needed for good wakefulness in the daytime, and about one in five adults believe they get at least one hour less than they need. With progressive age, the ability to tolerate shift work also declines.

CHANGES IN ADULT SLEEP

As one progresses from middle to older age, a number of changes occur in sleep. Among these is a tendency to go to bed and awaken at earlier times, a decline in stage N3 and sleep efficiency, and an increase in awakenings at night.

HOW AGING AFFECTS SLEEP

Sleep in older people

In adults over 65, a number of changes in sleep occur, though they sometimes become difficult to separate from the effects of illnesses (or medications). Visually scored stage N3 (slow-wave sleep) is very low and sometimes absent after age 65, particularly in males, the reasons for which are not understood. When the EEG is examined electronically, the amplitude of delta waves declines in older age. Older women may have less electronically measured delta activity, though not less visually scored slow-wave sleep, than men. The decline in slow-wave sleep may relate to the ease of being interrupted by noise, and differences in sleep-related growth hormone secretion (see pages 102–3). There may be some alteration in the normal relationship of frequency and amplitude of waveforms in the EEG. Typically, the slower the frequency, the higher the amplitude, but in the elderly this relationship may be diminished.

Another implication of declining slow-wave sleep in the elderly is that it may be related to a decreased ability to remove beta-amyloid proteins, toxic materials that accumulate in plaques in the brain in Alzheimer's disease. In slow-wave sleep, cortical neurons are particularly quiet, and it is thought that this is the time in which the brain can clean itself. In the normal elderly, those who have the least slow-wave sleep have higher levels of amyloid beta in their cerebrospinal fluid, in principle predisposing them to more deposition of plaques in the brain.

Slow-wave sleep is often used as a measure of the homeostatic drive in the two-process model of sleep-wake regulation (see pages 80–1), and it has been suggested that, in the elderly, the rate of decay of the homeostatic principle during sleep may decline. REM persists, variably but often with slightly lower amounts, but declines dramatically in the presence of some dementias. Perhaps the most clear change in the elderly is the increase in number of awakenings, both those of which the patient is aware, and of brief awakenings seen on the EEG. There is also a tendency in older people toward a circadian phase advance—for sleep to begin and end earlier, in effect the opposite of the tendency in adolescence. This may be compounded by the lifestyle in residential community living, as well as the diminished amount of sunlight exposure, often resulting in a decrease in the amplitude of the circadian rhythm of sleep and waking. One recommended remedy is to have bright lights available in the evening in community living facilities, to help make bedtimes later.

Sleep and memory processes

Among the outcomes of poor sleep in older people can be decrements in memory, attention, and ability to do tasks, which can be mistaken for dementia. We talked in Chapter Three about the role of sleep in processing memory. Processing of motor memories into longer-term memory during sleep may be associated with

MEASURING DAYTIME SLEEPINESS

An increase in daytime sleepiness can be seen using the Multiple Sleep Latency Test (MSLT). This measures physiological sleepiness by offering the patient four or five 20-minute nap opportunities across the daytime. The sleep latency (time from when the lights are turned out until sleep onset) is determined, and the average of the four or five sleep latencies is recorded (see pages 128–9). This increase in daytime sleepiness has been taken to mean that the decline in nocturnal sleep in the elderly is not due to a decreased need for sleep, but rather a decreased ability to sleep. The deteriorated sleep continuity is thought to be more important to decrements in daytime function than the decline in total nocturnal sleep. Because of increased sleeping in the daytime, the amount of total 24-hour sleep is variable and often the decline is relatively small.

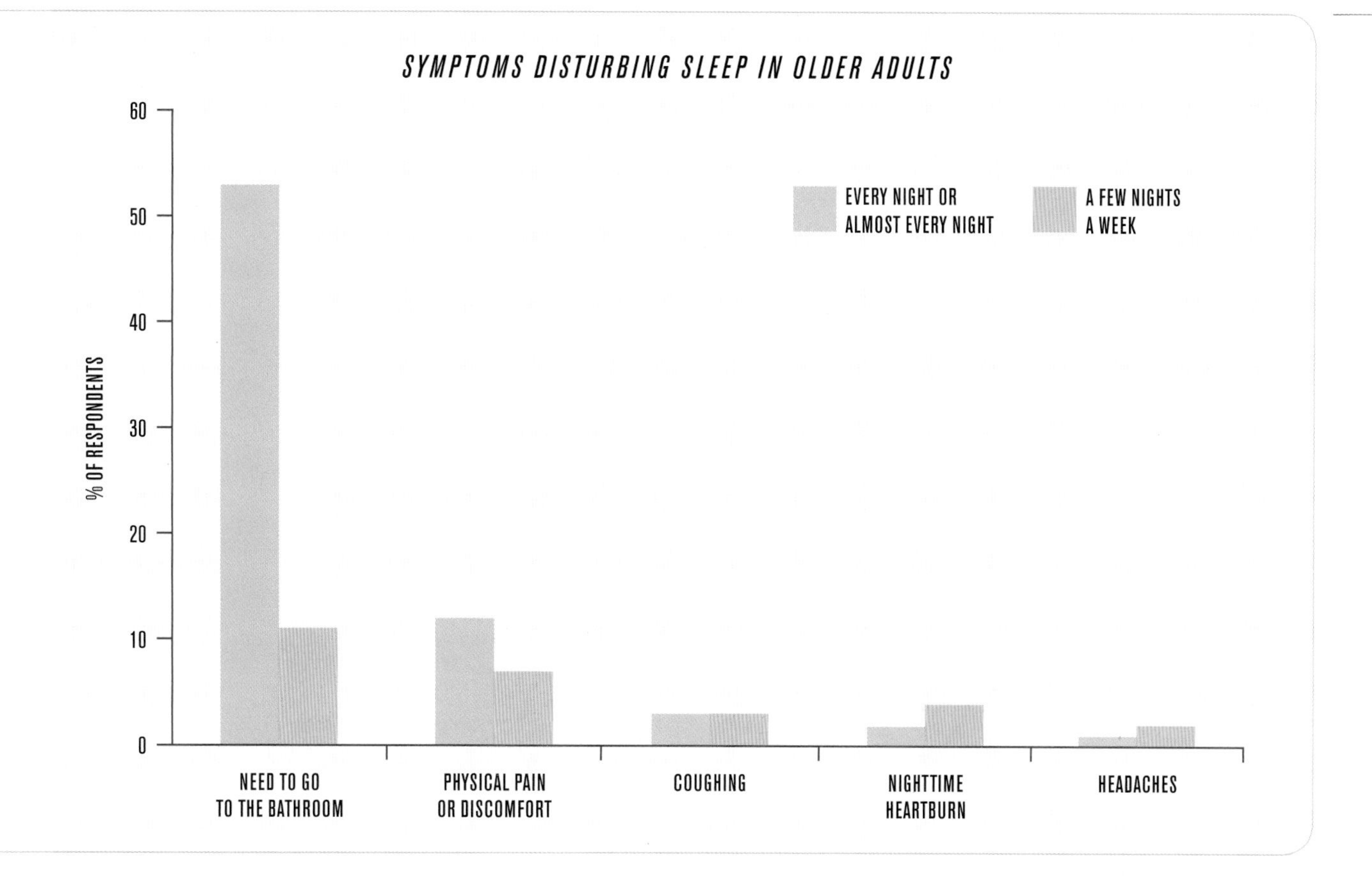

increased activity in the part of the brain known as the putamen as seen in fMRI studies, and in EEG spindle activity in daytime sleep. Both are relatively decreased in the elderly, in whom there is often less nocturnal sleep-related memory consolidation. Interestingly, in the elderly, though the overnight improvement in retaining procedural memory (memory of how to perform a particular action) is less marked, a nap after the training largely reduced these deficits.

It is difficult to make generalizations about the elderly, as the amount of individual variation greatly increases compared to younger adults. Moreover, a variety of sleep disorders including sleep apnea, medical illnesses, medications, and dementias have profound effects on sleep (See Chapter Seven, pages 108–43). The important thing to remember is that mild sleepiness is common, but severe sleepiness is likely to be associated with one of these conditions, and not normal aging. About 10–30 percent of elderly people describe marked sleepiness, which correlates with poorer social outcomes, productivity, and intimacy. Severe sleepiness, or a transition to more than 9 hours of night-time sleep, can be thought of as risk factors for current or later cognitive decline, and may be good indications for exploring medical or neurological causes of impairment. In summary, sleep changes in a number of ways in the elderly, particularly with a phase advance of the sleep–wake rhythm, a decline in slow-wave sleep, and an increase in awakenings. Mild daytime sleepiness is often seen, but severe sleepiness should alert one to the possibility of underlying illnesses.

NIGHT-TIME DISTURBANCES
In this National Sleep Foundation study of over 1,500 Americans aged 55-83, the most common causes attributed to sleep disturbance were the need to go to the bathroom and pain or physical discomfort. The need for the bathroom increased with age inside this population. Pain or physical discomfort were most common in those who slept the least, less than 6 hours per night.

Chapter Six

HORMONES AND SLEEP

Hormones are chemical messengers released by glands into the circulatory system, which help regulate a wide range of body processes such as growth, sexual development, the response to stress, and the transformation of ingested food into energy. The endocrine system is a general term for all the glands that secrete these hormones. In this chapter we will see how the functioning of the endocrine system has profound effects on how and when we sleep, and in turn that the secretion of hormones is influenced by sleep, circadian rhythms (Chapter Four), and other factors including age. As we discussed in Chapter Three, for instance, chronic deprivation of sleep can have major effects on hormones regulating appetite and blood sugar, and is associated with a tendency toward obesity. Thus sleep is necessary for appropriate endocrine function, and conversely the secretion of hormones influences the quality of our sleep.

WHAT DOES THE PITUITARY GLAND DO?

Endocrine secretion and sleep

Many hormones are secreted in periodic pulses and the particular pattern is often related to various aspects of sleep, including specific sleep stages, the REM–NREM cycle, or the overall timing of sleep/waking in the circadian day. There are many other influences as well, including whether it is light or dark, and even one's age. Often a hormone's secretion is affected by several of these factors at once, though to different degrees. Growth hormone is particularly interesting because studies of how drugs affect its release have shown that the same drug may have different—even opposite—effects when given during sleep or when awake during the day. The observation that a hormone's secretion is controlled by neurochemicals in a unique manner during sleep emphasizes that sleep is a very specific state with its own unique physiology.

The master gland

In this section we will focus on hormones of the pituitary gland, which is sometimes referred to as the "master gland," because the substances it releases into the blood have profound effects on a variety of glands around the body, such as the thyroid and adrenal glands, as well as having direct effects on tissues themselves. The pituitary is about the size of a pea, sitting in an opening in a bone at the base of the skull. It is connected by a narrow stalk to the bottom surface of the part of the brain known as the hypothalamus, just behind the optic chiasm (see pages 76–7). The anterior portion of the pituitary will be of most interest to us, as it is the source of a number of important hormones (see image opposite). The hypothalamus affects pituitary function by secreting its own releasing hormones, which signal the pituitary to secrete specific hormones. In general, the control mechanisms regulating the pituitary involve many of the same neurotransmitters that regulate sleep, such as serotonin, norepinephrine (noradrenaline), and acetylcholine (see pages 34–5).

METHODS OF STUDYING HORMONE SECRETION

Several kinds of studies can be done to determine whether and to what degree a hormone's secretion is related to sleep or circadian processes. The most basic is to draw blood samples for analysis at regular intervals around the clock. This gives a profile of how a hormone is secreted during normal conditions. Doing sleep EEG recordings during such a procedure also allows for determining whether secretion is related to a particular sleep stage. Another question that can arise is whether a hormone's secretion is related to the act of sleeping. This can be answered by drawing serial blood samples over 24 hours in a situation in which the sleep–waking pattern is reversed—having a person go to bed at 9:00 AM, for instance, instead of 9:00 PM. If a hormone which is normally secreted at night suddenly changes to morning secretion when the sleep–wake rhythm is reversed, the implication would be that its secretion is tied closely to sleep. Conversely, if upon reversing the sleep–wake cycle it continues to be secreted at night (even though the person's sleep period is now in the daytime), then the implication is that the hormone is regulated more by the time of the circadian day than by sleep. A third type of study is to keep a person awake continually for long periods of time and follow the secretion of a hormone, to see if it is related to sleep and waking. We will see examples of all these types of study as we look at secretory patterns of anterior pituitary hormones in the following sections.

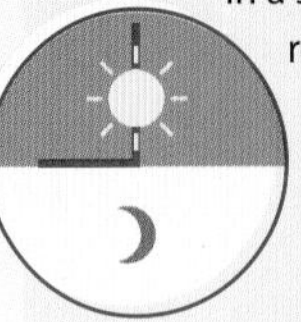

ENDOCRINE FUNCTION AND THE PITUITARY GLAND

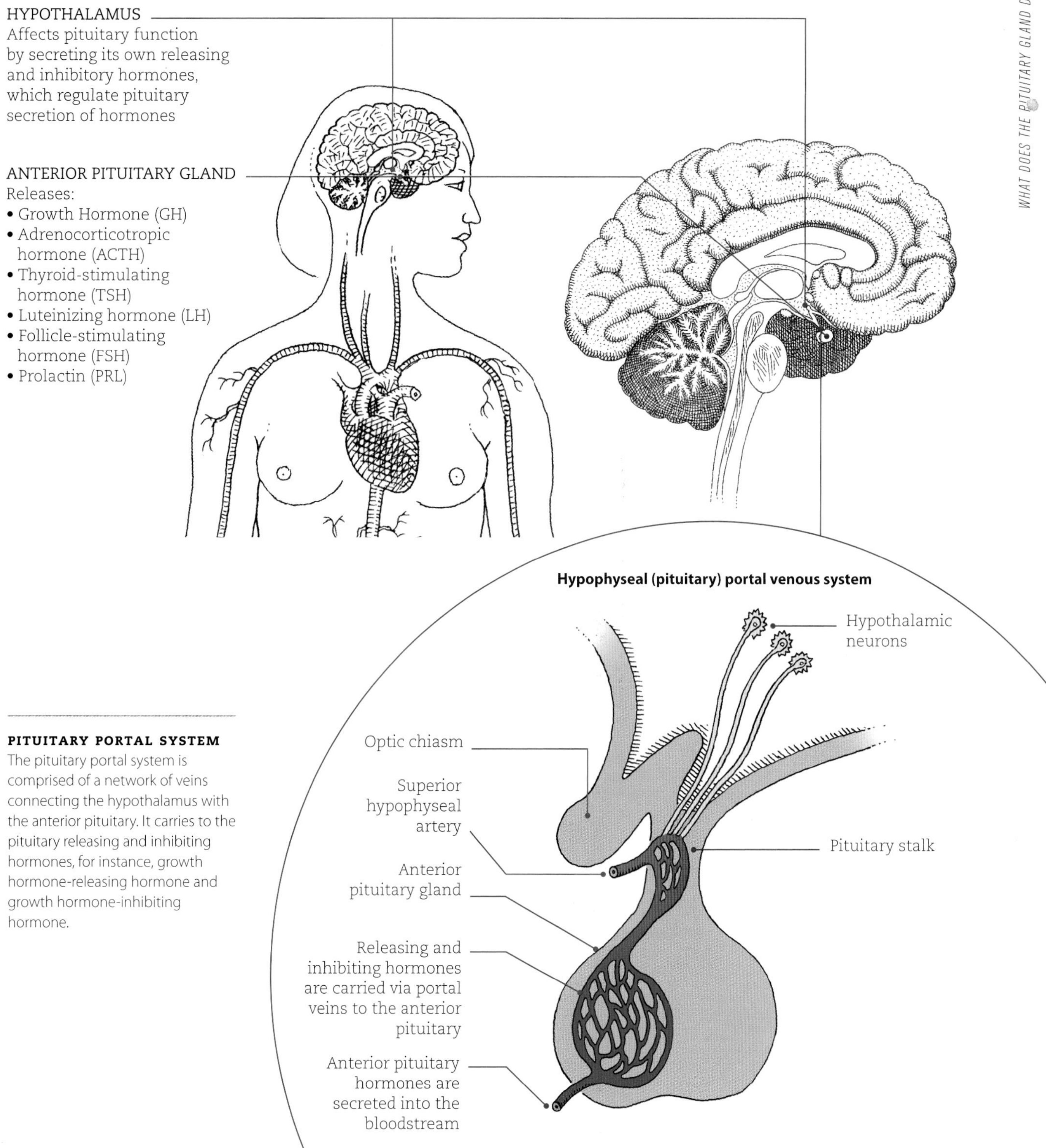

PITUITARY PORTAL SYSTEM

The pituitary portal system is comprised of a network of veins connecting the hypothalamus with the anterior pituitary. It carries to the pituitary releasing and inhibiting hormones, for instance, growth hormone-releasing hormone and growth hormone-inhibiting hormone.

HOW ARE CERTAIN HORMONES SECRETED DURING SLEEP?

The relation of sleep and circadian rhythms to hormone secretion

If we are to understand how hormones might influence sleep, and conversely how endocrine function in turn is influenced by sleep, a good starting point is to look at how the various hormones are typically secreted. The entire list of hormones is beyond the scope of this book, so we have selected specific hormones which are instructive about the relation of endocrine function and sleep.

Growth hormone: an exemplar of sleep-related secretion

Growth hormone (GH) is a peptide (a small protein), which is secreted by the pituitary gland, and which has a variety of functions. As its name indicates, it stimulates the growth of cartilage and long bones (until maturation at which time the latter are no longer capable of growth), as well as a variety of soft tissues and organs. Many of these effects are produced by peptides known as somatomedins, released by the liver in response to GH. It also has effects on metabolism of carbohydrates, protein, and lipids. GH tends to raise blood glucose levels in response to decreases induced by insulin. This property can be used as a stimulation test, by giving small doses of insulin and measuring the degree of the GH response. There is some evidence that GH may have behavioral effects as well, for instance after administration mice become more aggressive in certain situations.

In the late 1960s, landmark studies at Washington University in St. Louis found that in the first 90 minutes of sleep, a large secretory burst of GH occurred, lasting about 1.5–3.5 hours, and accounting for about 80 percent of the total 24-hour secretion. If sleep were delayed, the secretion occurred later as well. If a subject were awakened and kept up 2–3 hours, a new secretory burst occurred after returning to sleep. Almost half of secretion occurred during stage N3 (slow-wave sleep), although slow-wave sleep made up only 15 percent of the night. When sleep–wake 180-degree reversal studies were performed, GH secretion immediately moved to the new daytime sleep period. This indicated a clear association with sleep, and slow-wave sleep in particular, though later studies have found some circadian effects as well, such as a small secretory episode during nighttime sleep deprivation. Growth hormone release following injections of GH-releasing hormone is also subject to a circadian rhythm. The exact relation to slow-wave sleep is also complex, and some scientists think its release may be more related to the onset, rather than total amount, of slow-wave sleep.

Both daytime and night-time serum levels of GH decline in older adults. Sleep-related secretion is lower in older women than older men. It has been reported that in older women, there is a less clear relationship of sleep-related GH secretion to electronically measured delta wave activity (see pages 12–13), whereas the relationship persists in men. One possibility is that this gender difference is related to decreased amount of some measures of delta activity in women. Gamma hydroxybutyrate, the active compound in sodium oxybate, a medicine used in the treatment of narcolepsy (see page 129), has been reported to increase both slow-wave sleep and sleep-related growth hormone secretion in older adults. Nocturnal secretion of GH is also affected by disease states, and for instance is decreased in

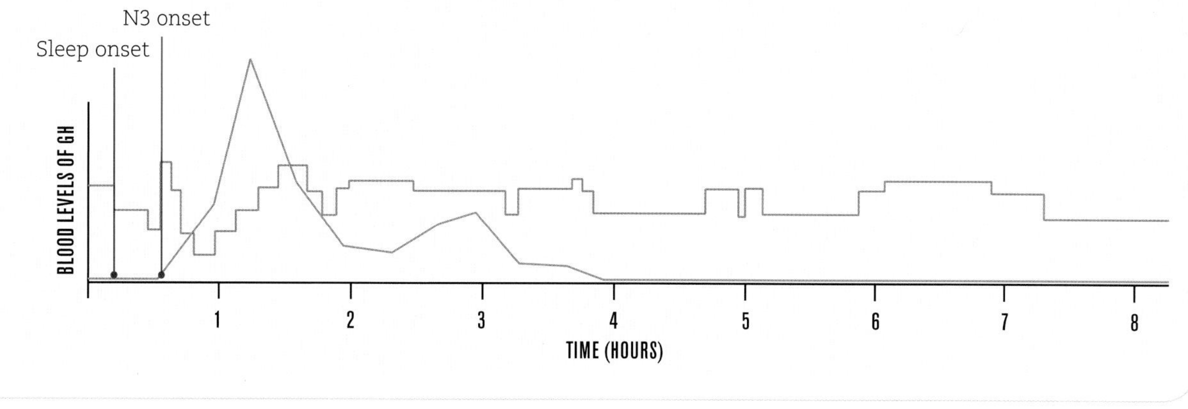

GROWTH HORMONE SECRETION ACROSS THE NIGHT
Typical pattern of growth hormone (GH) secretion in a young adult. In the first 90 minutes of sleep, there is a large secretory burst temporally associated with N3 (slow-wave sleep).

narcolepsy. One way of looking at the control mechanisms of hypothalamic regulation of GH is to observe the effects of giving drugs that modify the action of neurotransmitters related to sleep. Among these are acetylcholine and serotonin. In one study, volunteers were given methscopolamine, a drug that inhibits a major type of acetylcholine receptor (muscarinic receptors). It was found that methscopolamine had only very minor inhibitory effects on GH release when stimulated by insulin during the daytime; when given at night, however, sleep-related secretion of GH was almost totally abolished. This seemed to say that the role of acetylcholine in regulating GH release is very different in the two situations.

In order to explore the possible role of serotonin, the drug methysergide, which blocks serotonin receptors and in the past had been used for preventing migraine headaches, was given to normal volunteers. When methysergide-treated subjects received small doses of insulin during the daytime, there was a decreased GH response. This seemed to indicate that serotonin plays a facilitative role in GH secretion under these circumstances. When the methysergide-treated subjects slept at night, however, the opposite was found—their GH secretion was increased compared to those who had received placebo.

During sleep, then, serotonin appeared to have an inhibitory role on GH release. Altering the same neurotransmitter, therefore, could have opposite effect on GH, depending on whether it was given in the day or night. This was one of the first demonstrations that sleep appears to have its own unique physiology, very different from that seen in an awake subject in the daytime.

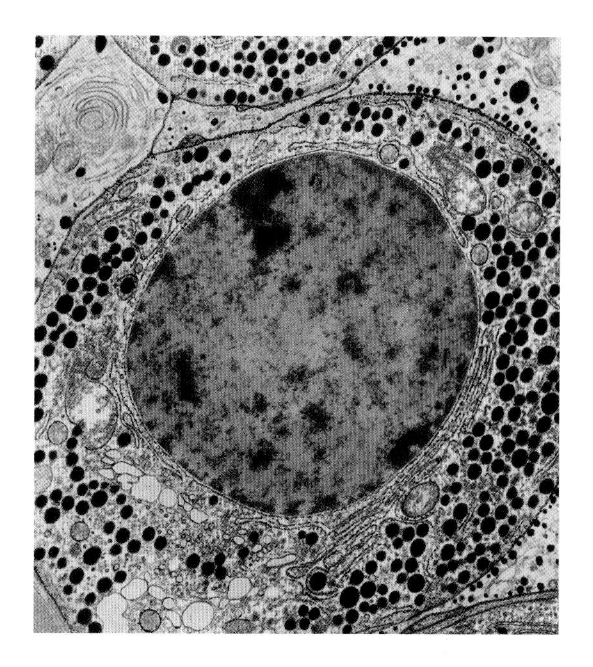

GH-PRODUCING CELL
This colored transmission electron micrograph shows a somatotropin, the type of cell in the anterior pituitary which produces growth hormone. The hormone itself is seen as the brown granules in the cytoplasm of the cell (yellow), the contents of the cell aside from the nucleus.

Cortisol and sleep

Cortisol is a steroid hormone released into the circulation from the adrenal glands that sit atop the kidney, in response to stimulation by adrenocorticotropic hormone (ACTH) from the anterior pituitary. The pituitary is in turn sensitive to amounts of cortisol in the blood, so that high levels reduce the amount of ACTH that is released. (This is known as "closed loop" regulation.) Among the actions of cortisol are increasing glucose production, decreasing the uptake of glucose into cells in most tissues of the body except the heart and brain, inhibiting the inflammatory response, and altering fat and protein metabolism. It plays roles both in providing energy stores during the part of the 24 hours when they are most needed, as well as in modulating the stress response.

Cortisol is secreted in a pulsatile manner, and the frequency and amplitude of these pulses are influenced by the circadian system. Average cortisol blood levels follow a predictable pattern around the clock, with rising levels about 4:00–8:00 AM. Upon awakening, there is often a spike of up to 50 percent, peaking about 30 minutes later; levels decline across the afternoon and evening, reaching a low point in the area of 1:00–3:00 AM. Cumulatively cortisol is released for about 6 hours over the 24, and about half of the 24-hour secretion is during sleep. In contrast to growth hormone, which is cleared from the blood very quickly, cortisol is removed relatively slowly, so it is hard to relate its secretion to a specific sleep stage, but it is often higher in REM sleep. Slow-wave sleep tends to be related to times of declining cortisol levels.

As with growth hormone, one can do studies of altering sleep in various ways to determine the relative importance of sleep and circadian processes in cortisol regulation. In Chapter Three we described studies in which volunteers were sleep deprived for long periods of time. When cortisol is studied in those conditions, its day–night pattern of secretion continues for up to 200 hours. When volunteers are kept in constant light for prolonged times, the secretory pattern remains relatively unchanged for up to three weeks. If people are put on a new sleep–wake schedule of very short (19-hour) or long (33-hour) days, cortisol secretion adjusts to the new schedule after 1–2 weeks. This relatively slow adaptation to changes in sleep is in sharp contrast to growth hormone, which changes immediately. Thus ACTH and cortisol are influenced by both processes,

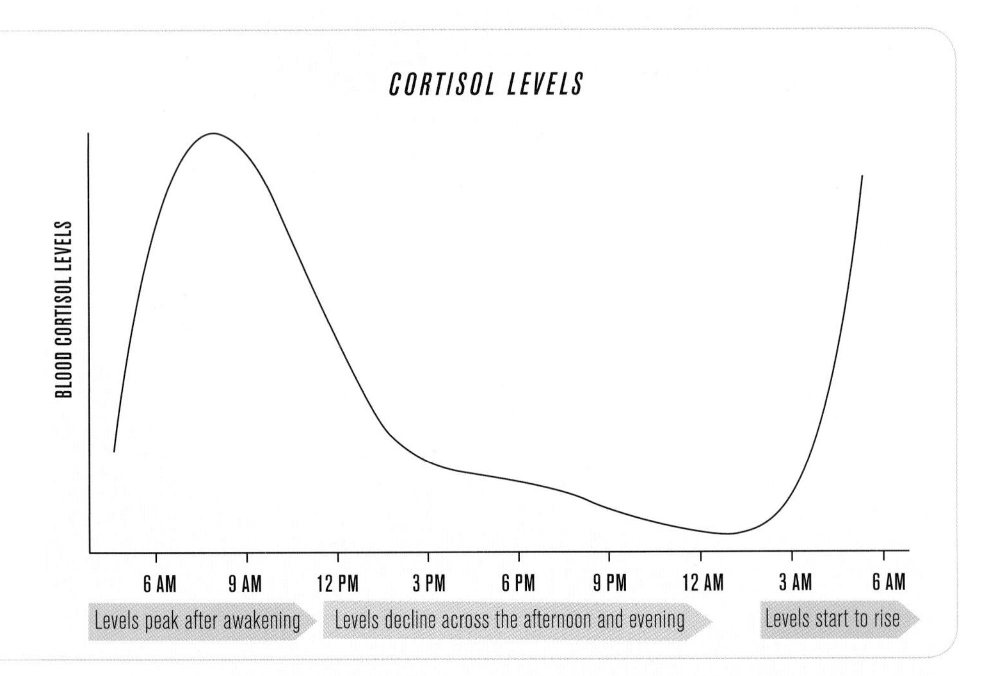

CORTISOL AROUND THE CLOCK
Most anterior pituitary hormones are secreted in multiple short bursts, or pulses. The amplitude and duration of individual pulses often varies throughout the circadian day, such as the increase of pulse frequency and amplitude of cortisol secretion in the early morning hours before awakening. Graphs which show the average hormone levels of groups of people tend to flatten out the individual pulses as they show average amounts of a hormone in the blood across time. The typical cortisol pattern includes rising blood levels in the early morning hours, with a large increase upon awakening, then declining throughout the day, with the lowest point around 1:00–3:00 AM.

CORTISOL FEEDBACK

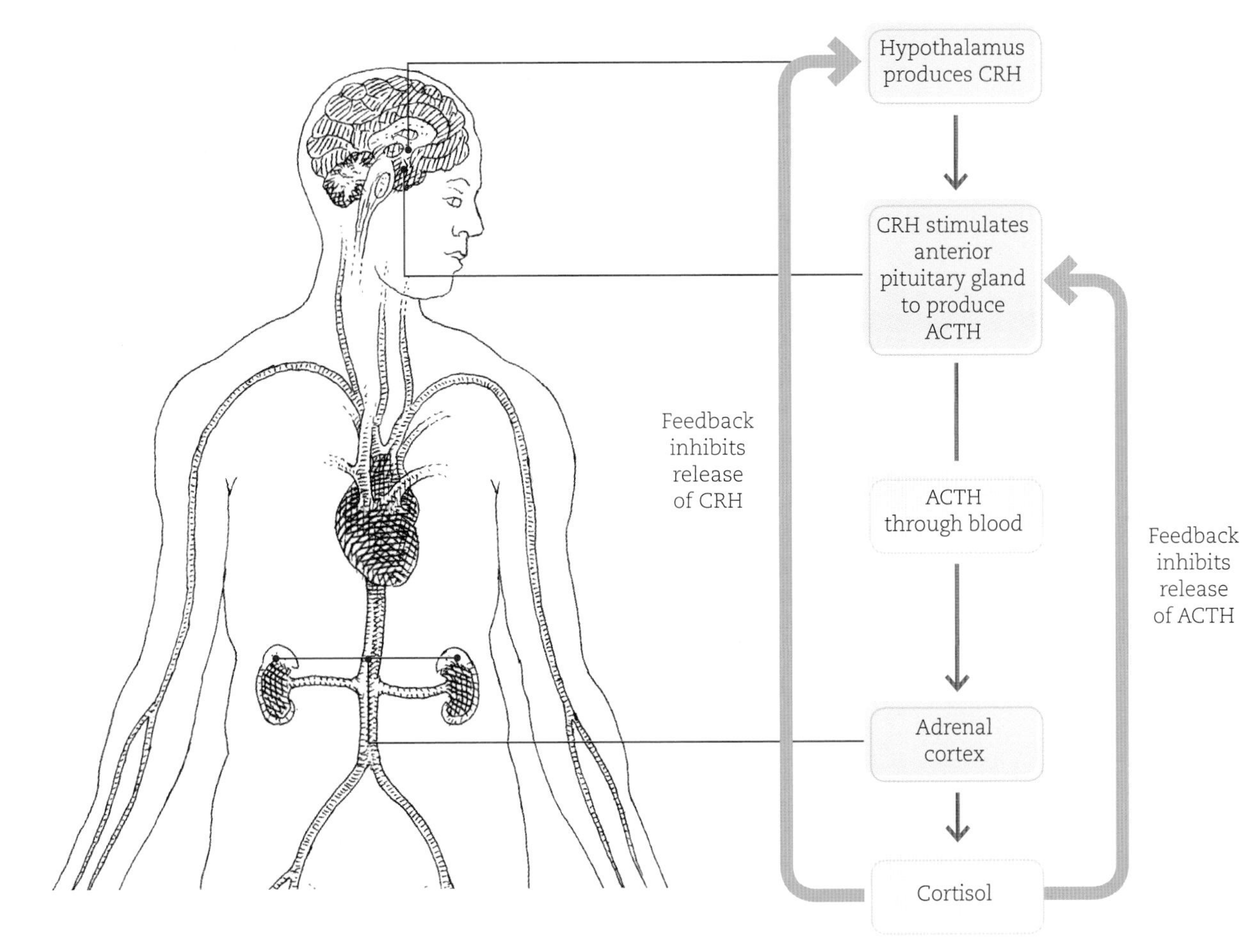

but circadian mechanisms play a much more important role compared to GH. Sleep deprivation may also affect cortisol release. After one night of partial or total sleep deprivation, cortisol levels on the following evening are elevated. The health consequences of higher cortisol levels are a subject of study, but as some evidence suggests effects on memory and ability to utilize the sugar glucose, it seems possible that chronic partial sleep deprivation might have deleterious effects. Total 24-hour cortisol secretion is higher in the elderly, and some researchers have felt this is related to their increase in nocturnal waking time and decline in amounts of REM sleep as discussed in Chapter Five. ACTH and cortisol secretion can also be altered in some sleep disorders. People with insomnia (see pages 146–7) may have higher levels of ACTH between 10:00 PM and 2:00 AM, and those with reduced total sleep time have been reported to have higher cortisol levels during those hours compared with those with more normal total sleep.

CLOSED LOOP REGULATION

The regulation of circulating cortisol levels involves what is known as a negative feedback closed loop system: The anterior pituitary releases ACTH into the circulation, which stimulates the release of cortisol from the cortex of the adrenal glands, located atop the kidneys. Circulating cortisol, in turn, influences the secretion of corticotropin-releasing hormone from the hypothalamus and modulates the amount of ACTH released in the anterior pituitary gland. The general principle of negative feedback closed loop systems is widely used in engineering control systems as well. For example, a house thermostat may be set to a particular temperature. This may cause the furnace to warm the room. The warmer air, in turn, may affect the thermostat, turning off the heat. This kind of control system leads to stability, and indeed an important function of the endocrine system is to help maintain stability of bodily processes (homeostasis).

Thyroid-stimulating hormone (TSH)

TSH is a glycoprotein (a protein linked to a carbohydrate) that enhances growth of the thyroid, a butterfly-shaped gland at the front of the neck which secretes hormones regulating metabolism (processing food to convert into energy and building blocks for other chemicals) throughout the body. TSH stimulates release of the thyroid hormones thyroxine and triiodothyronine, and these in turn increase the rate of metabolism. TSH is released from the anterior pituitary gland in response to thyroid-releasing hormone from the hypothalamus, and in turn is affected by negative feedback from circulating thyroid hormones. Plasma TSH levels rise in the evening before sleep onset and peak during the middle of the sleep hours. When sleep/waking are reversed by 180 degrees, there is an increase in the normal night-time secretion while subjects are kept awake. This seems to show that, in contrast to growth hormone, for instance, sleep is in some way inhibitory to TSH release.

Although acute sleep loss is associated with increases in TSH, after two weeks of partial sleep restriction this is not evident, suggesting that there is adaption of this particular process over time. Thyroid hormones also provide an example of how medical illnesses influence sleep. Patients with hyperthyroidism have increased amounts of slow-wave sleep, and conversely those with hypothyroidism have decreased amounts. Although there are many situations in which slow-wave sleep is decreased, there are relatively few in which it is increased. Generally, sleep doctors keep an eye out for the possibility of hyperthyroidism when they see a sleep study showing marked increases in slow-wave sleep.

Luteinizing hormone (LH)

LH is a glycoprotein released from the pituitary in response to gonadotropin hormone-releasing hormone (GnRH) from the hypothalamus, and which in turn is sensitive to both circadian influences as well

REGULATION OF THYROID FUNCTION

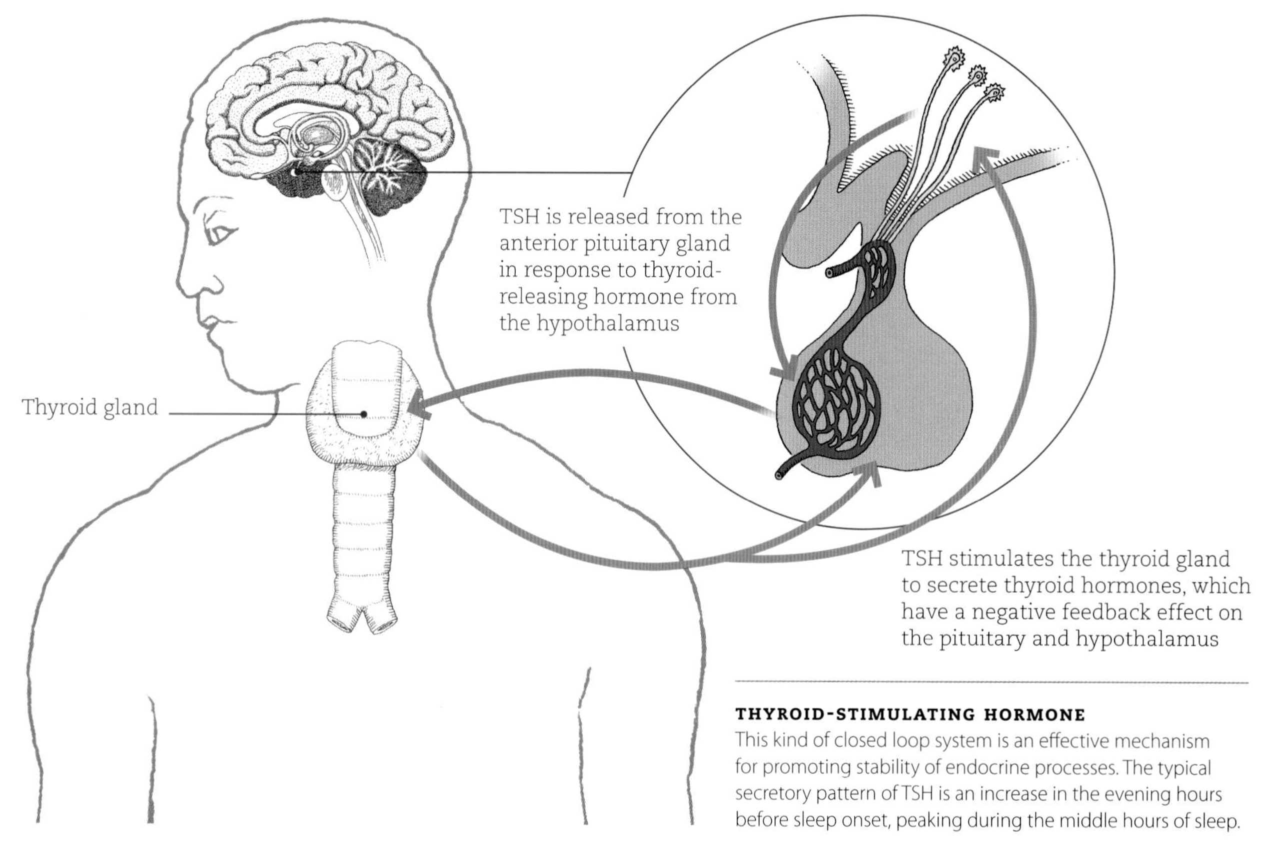

THYROID-STIMULATING HORMONE
This kind of closed loop system is an effective mechanism for promoting stability of endocrine processes. The typical secretory pattern of TSH is an increase in the evening hours before sleep onset, peaking during the middle hours of sleep.

as closed loop feedback from gonadal steroids. LH and the related hormone follicle-stimulating hormone (FSH) are gonadotropins (stimulating ovarian and testicular physiology). In children and adult men, there is no clear relationship of LH pulsatile secretion episodes to sleep. During puberty, however, in both sexes there is a sleep-related increase in pulses so that cumulatively there is substantially more secretion than during daytime waking. If sleep/waking are reversed 180 degrees, there is a shift to release during daytime sleep, though there still remains some night-time waking secretion, showing some circadian control as well. If the sleep period is delayed, LH secretion also is delayed in a similar manner and keeps its association with sleep. Toward the end of puberty, the amplitude of waking daytime secretory episodes goes up, so that in men there is often no longer a clear day–night difference. In women, LH secretion is modulated by the menstrual cycle and the 24-hour pattern varies with the phase of the cycle.

Other hormones and related endocrine topics

There are of course many other hormones that can also be looked at in terms of relationship to sleep and waking. In adult men, testosterone levels are low in the evening, increase early in sleep perhaps with an association to REM, then peak in the early morning hours. Prolactin, released from the pituitary, increases about an hour after sleep onset, with peaks at about 5:00–7:00 AM, and may be associated with REM sleep. Leptin, formed in fatty tissue and which promotes a sense of satiety, and grehlin, formed primarily in the stomach and which promotes a sense of hunger, as well as the broader issue of sleep deprivation predisposing to type 2 diabetes and obesity are presented in Chapter Three (see pages 64–5). The increased risk of pre-diabetes and diabetes in obstructive sleep apnea is presented in the discussion of this disorder in Chapter Seven (see pages 110–11).

Hormones influence a wide variety of body processes such as growth, energy production, and response to stress. Their secretion is often regulated by the same neurotransmitters involved in sleep. Studies of growth hormone have shown that the actions of these neurotransmitters may be very different in degree—or even opposite—depending on whether

GREHLIN AND LEPTIN

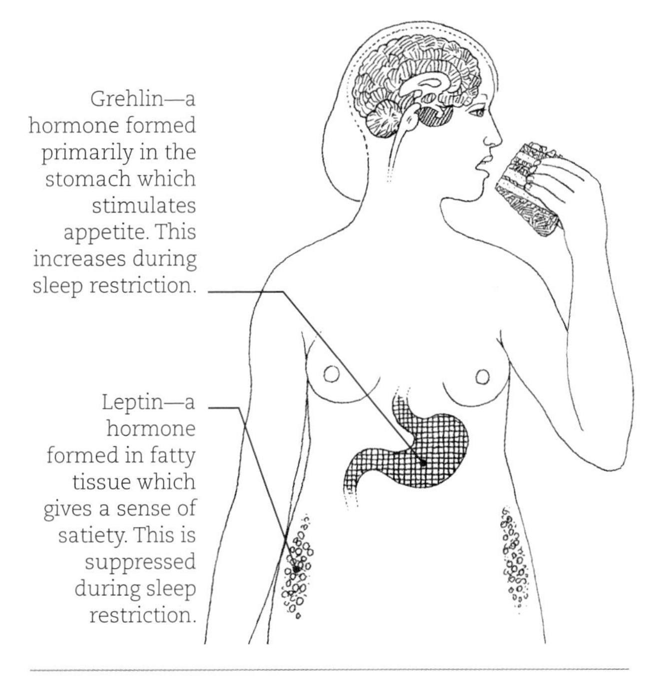

HORMONES AFFECTING APPETITE AND ENERGY USAGE
The regulation of food consumption is complex, including nervous system processes involving anticipation, reward, and energy homeostasis, as well as endocrine processes. Among the endocrine aspects are the hormones leptin and grehlin.

a person is awake or asleep. This is a good example of how physiology can be markedly different during sleep, something we have touched on, for instance, in the discussion of breathing responses to blood carbon dioxide levels during sleep (see page 27). Each hormone has a characteristic secretory pattern across the 24 hours, and some, such as growth hormone and prolactin, are primarily secreted in relationship to sleep. Others such as cortisol are relatively more heavily influenced by the time of the circadian day. TSH, at least in the short term, may be inhibited by sleep. LH and FSH are examples of hormones whose secretion in relation to sleep is most clearly associated with a specific phase of lifetime development (adolescence). There is a two-way relationship between sleep and endocrine function, each affecting the other. In some cases, inadequate amounts of sleep, as well as some sleep disorders, can alter endocrine function, for instance influencing appetite and energy utilization.

Chapter Seven

SLEEP DISORDERS

As described in Chapter One, sleep results from the harmonic working together of a wide variety of physiologic processes. It is not surprising that sometimes these complex systems malfunction, resulting in clinical sleep disorders. There are over 100 individual sleep disorders, but in general they are experienced by patients in one of three ways: insomnia, excessive sleepiness, or undesirable behaviors or experiences during sleep. In the last few decades there have been significant advances in understanding the physiologic abnormalities of, and developing treatments for, a wide range of sleep disorders. About 35–40 percent of the general population have difficulty going to sleep or daytime sleepiness. The underlying sleep disorders causing these problems have a significant impact on quality of life, and there are economic costs as well. In this chapter we will describe some of the most common sleep disorders, the symptoms they may produce, and how doctors diagnose them. Insomnia, which is the most prevalent sleep disorder, is dealt with in detail in Chapter Eight. It is important to bear in mind that these chapters are designed to provide information about a variety of sleep disorders and their treatment, to aid the reader's understanding when seeking care from a licensed medical practitioner.

SLEEP-DISORDERED BREATHING

What are obstructive and central sleep apnea?

Sleep-disordered breathing is a good example of the unique physiology of sleep: in general it represents the situation of a person having normal respiration while awake, but then having cessations of breathing when asleep. These gaps in breathing, known as apneas (or in the case of partial gaps, "hypopneas") are of two general types:

1. Obstructive events: in which there is cessation of airflow due to obstruction in the upper airway. The obstruction is caused by the throat muscles relaxing, eventually closing the throat for short periods. (There is often a predisposition for obstruction due to having a narrower airway for a variety of reasons.) During these episodes, the muscles that promote breathing (the diaphragm and muscles in the chest wall) continue to work.

2. Central events: in which the brain periodically fails to send signals to the body to breathe. During these episodes there is no movement of the diaphragm or chest wall musculature. In addition, there is a condition in which a person has apneas of mixed type, exhibiting both central and obstructive qualities.

As a consequence of these gaps in breathing, the amount of oxygen in the blood drops and the amount of carbon dioxide rises, often resulting in brief awakenings throughout the night. Ultimately, a constellation of symptoms may appear, including insomnia, excessive sleepiness, fatigue, sexual difficulty, and hypertension.

Obstructive sleep apnea (OSA) is most common in middle age, in which at least in its milder forms can occur in as many as 9 percent of women and 24 percent of men. After age 65, rates for men and women become more similar, except in women taking hormone replacement therapy. The frequency is higher among the obese, and those with adult-onset diabetes and high blood pressure. It also occurs in 1–5 percent of children.

Central sleep apnea is more common in middle-aged and older adults; most, but not all studies have found it more common in men. There is also a form of central sleep apnea known as Cheyne–Stokes respiration, in which apneas appear in the context of repeated patterns of gradually increasing and decreasing respiration. This typically occurs in patients with congestive heart failure, in which it might appear as frequently as in 45 percent of cases. Sleep-disordered breathing can also appear as a worsening of respiration during sleep among people who have pulmonary disorders such as chronic obstructive pulmonary disease (COPD). There is also a group of sleep-related disorders known as sleep-related hypoventilation disorders. A prominent

TERMS USED IN THIS CHAPTER

Central: As in central hypersomnia or central apnea, this refers to the central nervous system (the brain and spinal column). Thus central sleep apnea refers to a kind of sleep apnea in which the brain periodically fails to send messages to breathe to the diaphragm.

Hypersomnia: Excessive sleepiness. One category of illnesses in this chapter, for instance, is "central hypersomnias," which refer to disorders in which sleepiness is a result of malfunction of the central nervous system.

Idiopathic: When an illness is called idiopathic, it means that the cause is not known. This is the case, for instance, for "idiopathic hypersomnia," a type of insomnia that is not well understood.

Parasomnias: This refers to disorders involving undesired or unpleasant experiences during sleep, such as sleepwalking or night terrors.

OBSTRUCTIVE SLEEP APNEA

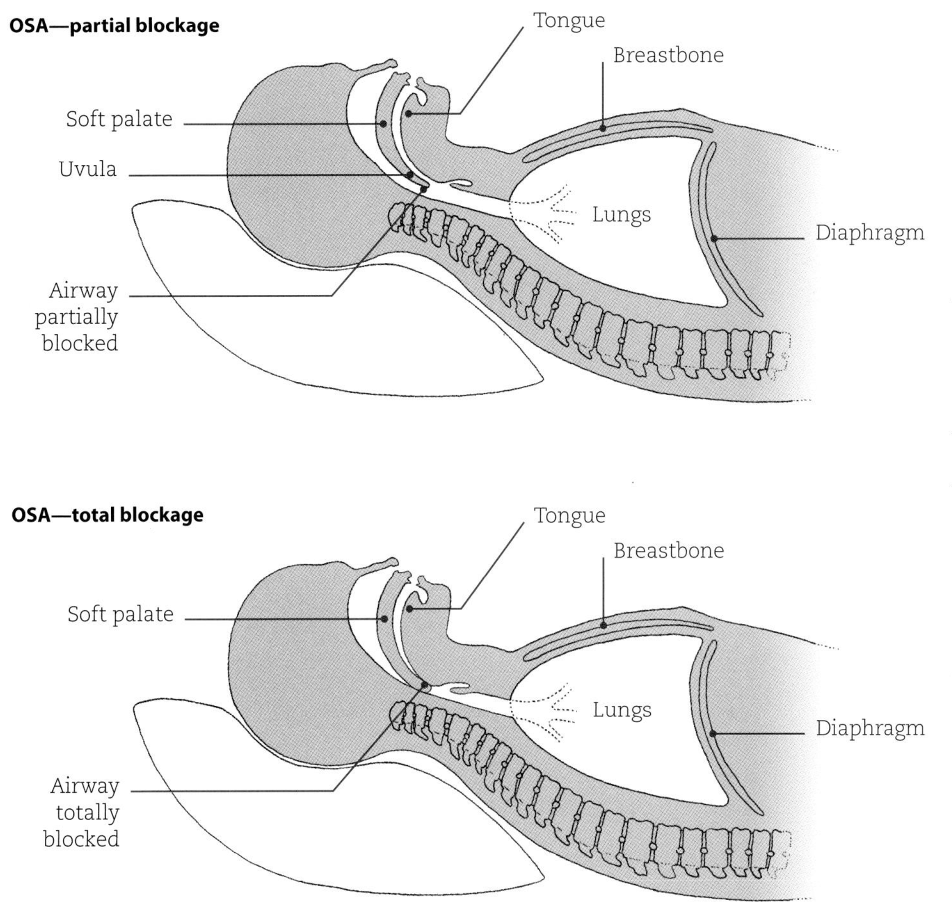

OBSTRUCTIVE EVENTS
In obstructive sleep apnea, during sleep the upper airway becomes blocked at various possible points. The pharynx does not have a bony structure to help keep it open. There are muscles which help to do this, but their activity is decreased during sleep. A narrow airway, which predisposes to blockage, is often present. This may be due to genetics, obesity, illness (for instance, a large tongue in hypothyroidism), or other factors. Lying on one's back also predisposes to airway collapse, and one recommendation often made for people with mild to moderate OSA is to avoid sleeping in that position.

example of these is the obesity hypoventilation syndrome, formerly known as Pickwickian syndrome (after a character in Charles Dickens' *Pickwick Papers*). Dickens described him this way: "Joe was a wonderfully fat boy, standing upright with his eyes closed." The syndrome includes significant obesity, severe sleepiness, and decreased breathing when awake, in the absence of obvious pulmonary disease, with increases in blood carbon dioxide levels, and usually obstructive sleep apnea. One possible cause being studied is that it might involve a decreased sensitivity to the hormone leptin (see pages 64 and 107), which among other actions may increase respiration in response to rising blood carbon dioxide levels.

Snoring

At the milder end of the spectrum of difficulties with breathing during sleep is snoring, which may occur in 32 percent of adults and 7 percent of children. It is prominent during inspiration (breathing in), and most frequent when one is lying on one's back. Snoring is a characteristic symptom of OSA, but the majority of people with chronic snoring do not have the full-blown syndrome. On the other hand, there is increasing evidence that chronic snoring is not completely benign (aside from the distress that it may cause one's bed partner), but may be associated with health consequences including an increased risk of vascular disease and stroke.

DIAGNOSING SLEEP APNEA: THE POLYSOMNOGRAM

How does a sleep study detect sleep apnea?

In Chapter One, we discussed the techniques of basic sleep recordings, in which the electroencephalogram (EEG), electromyogram (EMG), and electrooculogram (EOG) are used to determine sleep stages. As the field of sleep medicine evolved, a number of other recording channels were developed to assess additional aspects of physiology. The technique of performing such a study with this whole array of channels is referred to as polysomnography, and the resultant output is known as a polysomnogram (PSG). Among these additional channels of information found in a typical PSG are:

- **Nasal airflow:** sensors that measure air temperature (the air you breathe out is warmer than the air you breathe in) or changes in air pressure in the nose give an estimate of ventilation. Some laboratories also have a special electrode near the nose that measures the amount of carbon dioxide in the air that is breathed out.
- **Respiratory effort:** bands are placed around the abdomen and chest wall to measure the effort of the diaphragm and chest wall muscles respectively. This can be done by measuring the degree of stretching of an elastic band (strain gauges) or

A TYPICAL PSG SAMPLE WITH NORMAL RESPIRATION

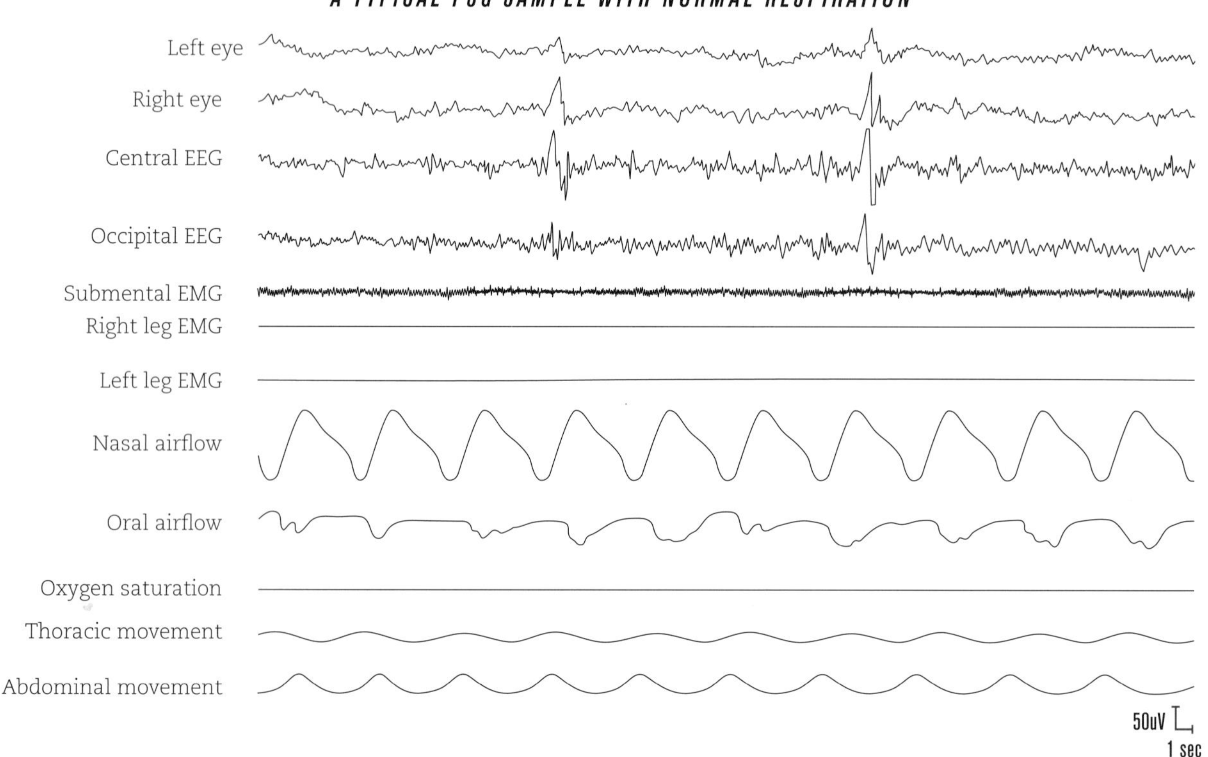

with a more sophisticated technique known as "inductive plethysmography." As we mentioned earlier, the distinction between obstructive and central apneas depends on the presence (obstructive) or absence (central) of respiratory effect during periods of absence of nasal airflow.

- **Arterial oxygen saturation:** In order to determine whether cessation in respiration results in decreases in the amount of oxygen in the blood, a device known as an oximeter is attached to a finger. Basically it detects the color of hemoglobin in the blood, which reflects the amount of oxygen it is carrying. This number is expressed as a percentage of the total potential oxygen-carrying capacity of hemoglobin and is referred to as arterial oxygen saturation.
- **Cardiac rhythm:** A cardiac channel measures the heart rate and is used to assess the possible presence of abnormal heart rhythms which may occur during apneas.
- **Leg electromyogram:** Pairs of EMG electrodes are placed over the anterior tibialis muscles (on the shin of the lower legs), in order to detect the leg movements of periodic leg movement disorder (see pages 122–3).

On the polysomnogram, apneas are defined as a decrease in oronasal air flow of 90 percent or more, lasting at least 10 seconds. A partial apnea, known as a hypopnea, is present when nasal airflow is decreased by at least 30 percent, accompanied by a drop in arterial oxygen saturation of at least 4 percent. These respiratory events are considered obstructive if they occur in the presence of respiratory effort seen in the chest or abdominal channels; they are considered central when respiratory effort is absent. Some respiratory events can be of mixed type, beginning as a central event, then showing respiratory effort as it progresses. The number of respiratory events is calculated,

POLYSOMNOGRAM

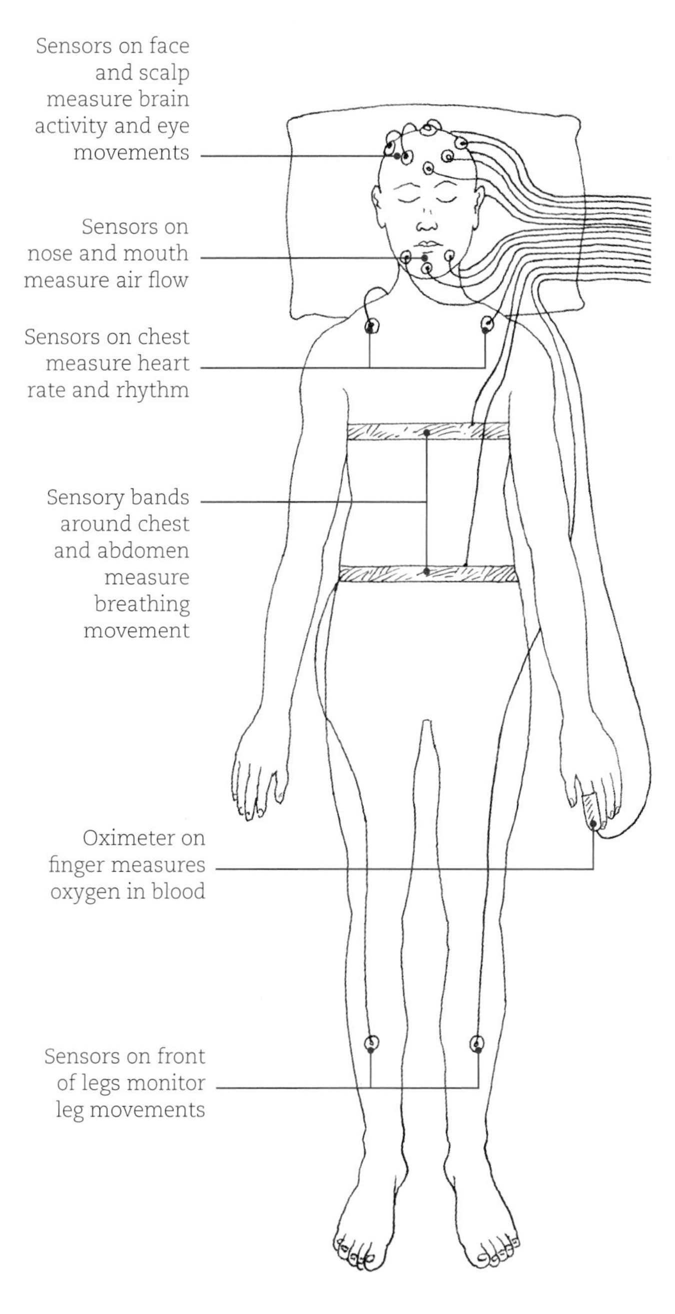

MEASURING PHYSIOLOGICAL ACTIVITY (LEFT AND ABOVE)

The "poly" in "polysomnogram" emphasizes that a wide range of physiologic processes are measured during sleep. Among these are the EEG, EOG, and submental EMG (see pages 22–3); heart rhythm; nasal airflow; arterial oxygen saturation; movement of chest and abdominal wall during respiration; and movement of the legs. Many of these physiologic processes are altered at the same time in sleep disorders. For instance, during an obstructive event in obstructive sleep apnea, nasal airflow ceases, there is a drop in arterial blood oxygen, heart rhythm may be altered, and respiratory muscle activity may increase in an effort to move air past the blockage.

RECORDING APNEAS

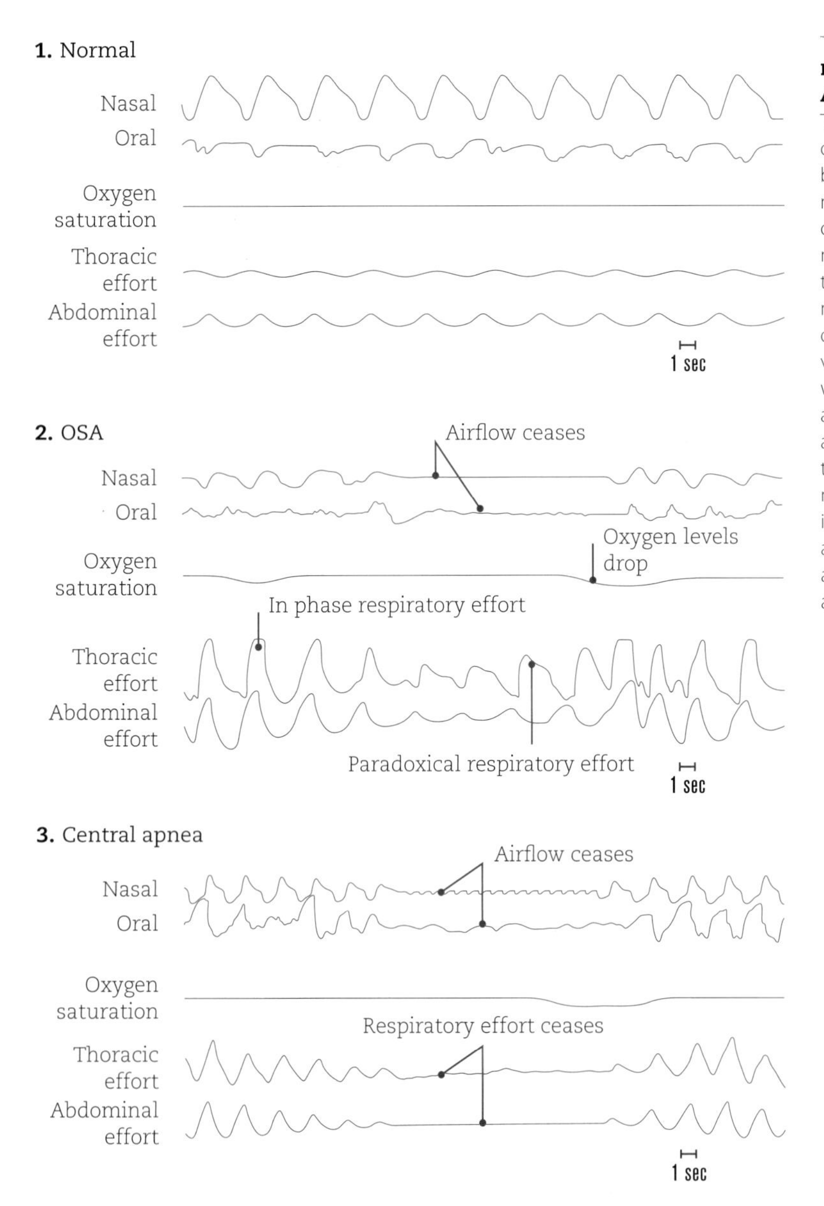

BREATHING DURING NORMAL SLEEP AND IN SLEEP APNEA
The channels of a PSG related to breathing and blood oxygen when studying sleep apnea. 1. Normal breathing during sleep—there is airflow through the nose and some through the mouth, good arterial oxygen saturation, and the chest wall and diaphragm move in synchrony. 2. During an obstructive event, there is a blockage of the pharynx. Airflow from the mouth and nose cease, and there is a drop in arterial oxygen saturation. The chest wall and diaphragm work very hard to move air past the obstruction, sometimes working so hard that they lose their coordination, a condition known as paradoxical effort. Ultimately a reflex response to changes in blood chemistry triggers a brief arousal, then a return to sleep with normal ventilation. 3. In central sleep apnea, there is a cessation of nasal and oral airflow and a drop in arterial oxygen saturation, but in this case, there is also cessation of respiratory effort by chest muscles and diaphragm.

DAYTIME CONSEQUENCES OF OBSTRUCTIVE SLEEP APNEA

- Daytime sleepiness
- Increased rate of accidents in vehicles or at work
- Male sexual dysfunction
- Mood changes
- Difficulties with memory and concentration
- Decreased overall quality of life reported on quality of life scales

and is often expressed as the number of apneas and hypopneas per hour of sleep (the "apnea/hypopnea index," or AHI). In general, sleep apnea is diagnosed when the AHI is greater than five per hour.

In the early years of sleep medicine, polysomnograms were performed primarily in the setting of the sleep laboratory. As technology became more sophisticated, devices became available for recordings at home. Their capability varies. Most commonly they measure air flow from the nose, respiratory movement of the chest wall and diaphragm, blood oxygen and heart rate, often without measures of EEG sleep. Periodically, guidelines for use of home testing have become available, and the standards for their use are still evolving. In general, home testing is more effective for ruling in a diagnosis of sleep apnea (that is, confirming a strong clinical suspicion of its presence), but less effective in ruling out the presence of sleep-disordered breathing. Sleep laboratory studies continue to play an important role in a number of situations, such as in following up a negative home study when there remains a strong suspicion of sleep apnea, or when treatment based on a home study is not effective.

Symptoms and consequences of obstructive sleep apnea

People with obstructive sleep apnea (OSA) often complain of daytime sleepiness and fatigue, decreased sexual function, decreased motivation, memory trouble, and mood changes. There is often a history of snoring, morning headaches, or of a bed partner having witnessed periods of cessation of breathing. They are frequently overweight, and have larger neck size. Indeed people with a neck circumference of 19.2 inches are twenty times more likely to have obstructive sleep apnea than those with smaller necks. People with central sleep apnea are more likely to complain of insomnia and awakenings during the night, though some also describe daytime sleepiness.

OSA is associated with a variety of problems in one's daytime life. People with OSA are two to thirteen times more likely to be involved in automobile accidents. Although this is usually associated with sleepiness, they may also have more accidents not related to sleepiness as well. The accident rate may be compounded by sleepiness due to work schedules, particularly in long-haul truck drivers. About half of male OSA patients report histories of occupational accidents of various types. People with OSA often describe difficulty doing mental tasks and getting work done, particularly monotonous assignments. On formal psychological testing they tend to show impaired vigilance and ability to sustain attention, slowing and increased errors on tests of cognitive processing, decreased ability to absorb, store, and bring back information on memory tasks, and difficulty in judgment as well as in making plans and carrying them out. Many people with OSA also report decreased quality of life on a variety of measures.

MEDICAL CONSEQUENCES OF OBSTRUCTIVE SLEEP APNEA

OSA is associated with a variety of medical conditions, including increased risk of:

- Hypertension throughout the body, or in the lungs (pulmonary hypertension)
- Stroke
- Myocardial infarction
- Sudden death
- Death from cancers
- Diabetes mellitus
- Changes in white matter and gray matter in the brain

OSA is associated with a variety of medical disorders. People with OSA are two to three times more likely to have or develop systemic hypertension (high blood pressure throughout the body), which can only be partially accounted for by their weight, sex, or age. In general, treatment with continuous positive airway pressure (see pages 116–17) reduces blood pressure to some degree. They are also more likely to develop high blood pressure inside the pulmonary artery system ("pulmonary hypertension"). There is a proportionately higher frequency of coronary artery disease, and a higher rate of death from coronary artery disease during the night-time hours compared to the general population. The incidence of stroke and death is higher as well, even after allowing for factors such as high blood pressure, diabetes, and smoking. People with significant OSA are more likely to develop a pre-diabetic state (in which blood insulin levels are higher than normal, but not high enough for a formal diagnosis or diabetes), or true diabetes. Although there are some hopeful data, it has not yet been fully determined whether in the long-term positive airway treatment of OSA will stop or slow down the progression from pre-diabetes to diabetes.

In children, OSA may be associated with overactivity, attention or cognitive difficulties, eneuresis (bed-wetting), sleepiness, and growth difficulties. Some animal research has suggested that the peptide orexin/hypocretin (see pages 36–7) may be involved in respiratory-induced disturbance of sleep and growth, which were experimentally improved by the orexin antagonist almorexant.

TREATMENT OF SLEEP-DISORDERED BREATHING

What can be done to help people with sleep apnea?

For obstructive sleep apnea, weight loss is an important long-term step, in conjunction with more specific treatments described below. Because snoring and obstructive apneas are usually worse when lying on one's back, it is good to learn to sleep on one's side instead. There are T-shirts available with pockets in the back for a tennis ball, which make it uncomfortable to lie on your back, as well as pillows designed for the same purpose. Since both alcohol and many sedatives (see page 158) have varying degrees of respiratory suppressant qualities, they should be avoided. There are also some medical steps which can be carried out by a physician, including checking for enlarged tonsils or nasal obstruction, and sometimes nasal steroids are used. Nasal dilators made of adhesive to hold the nose open have had mixed results.

Treatment of obstructive sleep apnea

The conservative steps of weight loss, avoidance of alcohol and sedatives, and training to sleep in the lateral position that we just mentioned are the starting points with OSA. Perhaps the most common treatment involves the use of positive air pressure devices. In cases of very significant obesity, surgery for weight loss is sometimes employed.

Positive airway pressure (PAP)

Positive airway pressure is generally the treatment of choice for obstructive sleep apnea. Introduced for this particular use in the 1980s, in its simplest form it can be thought of as a fan in a box, with a hose leading to a flexible mask placed over the nose. Air under low pressure is blown into the upper airway, and acts as a "pneumatic splint" to make it larger. Originally it was thought that it might also cause airway muscles to reflexively open the airway further, but this is now less clear. Some patients feel discomfort breathing air outward (expiring) when there is a low pressure blowing in. For this reason, there are also devices which blow air at a higher pressure when one breathes in, and a lower pressure when breathing out. There are also devices that automatically sense the pressure needs for the patient, and are capable of altering the pressure during the night. Though this might appear to have a number of advantages, the data at this point indicate that although they provide a slightly lower average pressure and might be more comfortable, there is no difference in the number of hours per night that patients use these devices. Used as directed, PAP greatly reduces the number of apneas and improves arterial oxygen saturation, reduces sleepiness, and improves blood pressure in patients with moderate to severe OSA. It has been shown to improve measures of quality of life in patients with moderate to severe OSA, though it does not fully reverse deficits in attention and cognitive function. When people with OSA have depression as well, there are some data to indicate that the depression is more responsive to treatment when the patient is receiving PAP. The effects of PAP on driving safety have also shown its benefits. A study of truckers with OSA found that those who faithfully used their auto-adjusting positive airway pressure devices had rates of preventable accidents similar to drivers without OSA, while those who were non-compliant with the use of positive air pressure had five times the number of accidents.

Aside from the discomfort of breathing out against pressure, most other side effects of PAP are related to discomfort of the mask or air leaks around it.

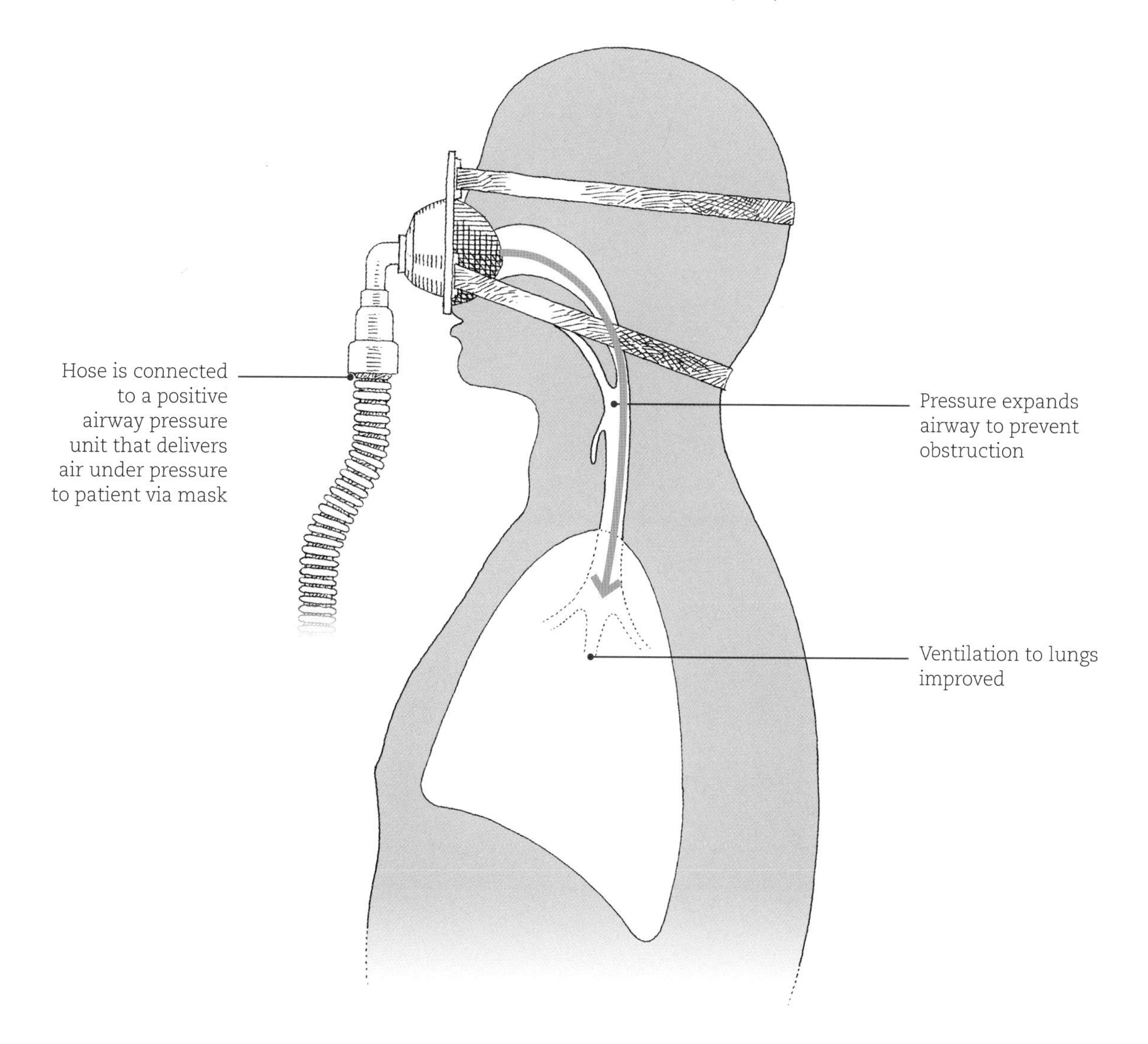

For people who feel uncomfortable with a particular mask, a variety of masks are available, and there are also soft foam-cushioned flexible tubes that insert into the nostrils. There can be nasal congestion, or irritation of the conjunctiva of the eyes from air leaks, and skin irritation from the mask. Sometimes warming and humidifying the air can be helpful, particularly in patients with nasal congestion. Overall, the side-effect profile is very benign compared to surgical alternatives, and it is considered the treatment of choice for OSA and some forms of central sleep apnea.

PAP TREATMENT
Illustration of a positive airway pressure mask placed over the nose in an obstructive sleep apnea patient. Masks come in a variety of shapes and sizes. The PAP apparatus itself comes in a variety of levels of sophistication, including those with a simple pre-set level of pressure, some with pressure which differs between inspiration and expiration, and others which automatically adjust to the amount of pressure needed.

Surgical approaches and implanted devices

Surgical treatments for OSA include tracheotomy and surgical procedures, which include the uvulopalatopharyngoplasty (UPPP). When sleep apnea became widely recognized clinically in the early 1980s, tracheostomy (the insertion of a breathing tube into the airway through the neck) was the only treatment available. Although very effective, it was not a very satisfactory solution because of its nature as well as medical complications, and was largely replaced by the UPPP. The latter involves taking out or repositioning tissue obstructing the airway in the nose, palate, and base of the tongue. It has the advantage of often being a one-time treatment, in contrast to PAP, which needs to be used night after night. On the other hand, it is a major kind of surgery with its own potential risks, such as changes in the voice or regurgitation of fluids into the nose. Many sleep doctors reserve surgery for patients who for one reason or another have not had success using PAP or prefer not to use it, or for patients with severe illness with more than twenty apneas or hypopneas per hour, or those with abnormalities in the anatomy of the airway. Although it is often quite helpful, some UPPP patients continue to have significant sleep-disordered breathing and will additionally need PAP. Depending on the anatomy of a specific person's airway, there are also alternative and more specific surgical techniques, often done by endoscopy under sedation.

Hypoglossal nerve stimulation

Another approach has been electrical stimulation of muscles to open the airway. Hypoglossal nerve stimulation (HNS) involves the implantation of a device in the chest wall, from which run wires to a breathing sensor and to the hypoglossal nerve. When one is breathing in, the hypoglossal nerve is stimulated, leading to a contraction of the genioglossus muscle (which connects the chin and tongue), pulling the tongue forward. It may have some benefits in some patients unable to tolerate PAP, but also has its own side effects, including an uncomfortable feeling, dry mouth, effects on nerves, and possible abrasions. It is only used by specialized laboratories at this time, and is generally considered not yet to be a first-line therapy for OSA.

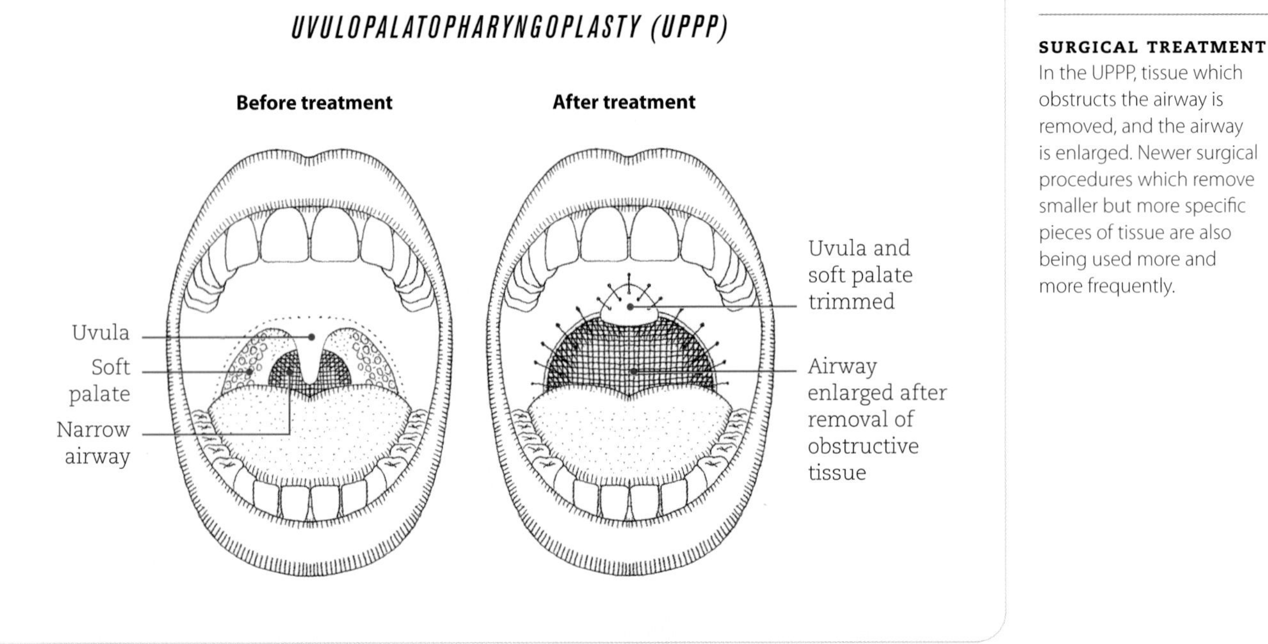

SURGICAL TREATMENT
In the UPPP, tissue which obstructs the airway is removed, and the airway is enlarged. Newer surgical procedures which remove smaller but more specific pieces of tissue are also being used more and more frequently.

Medications
Some medications have been used by doctors for mild OSA, with limited success. These include some tricyclic antidepressants, the serotonin reuptake inhibitor fluoxetine, and the mixed serotonin receptor blocker mirtazapine. Their benefits are limited, and none are approved in the United States or Europe specifically for this purpose. In contrast to these medicines, which were given with the goal of decreasing the sleep-disordered breathing itself, is the drug modafinil, which is given to combat the sleepiness associated with OSA. It appears to be reasonably effective at this. One potential limitation is that some patients receiving it feel better and then decrease their use of PAP, so it needs to be emphasized that continuing PAP to address the sleep-disordered breathing itself is very important.

Oral appliances
Oral appliances, which have a history going back to the nineteenth century as treatments for snoring, are various in form but generally are designed to fit in the mouth and to open the airway. This is most commonly done by mechanically advancing the mandible (lower jaw) or attaching to the tongue with a suction cup and pulling it forward. There is at least one good study showing that bed partners of people with non-apneic snoring report a decrease in severity of snoring when these devices are used. After two years or so, only about two-thirds or less of people continue to use them, often dropping out due to lack of effectiveness. There are some data indicating that they can improve sleepiness, breathing, and blood oxygen in mild to moderate sleep apnea, though to a lesser degree than PAP. Although there is very active work in this area, in the United States they generally have not yet been formally approved for treatment of OSA.

Treatment of central sleep apnea
Treatments for central sleep apnea in the absence of pulmonary disease and increased levels of blood carbon dioxide are less clear. PAP seems to help many patients. One rationale is that there may be initial closing of the airway which triggers a reflex in which the brain's signal to breathe is inhibited. In heart-failure patients who have not had success with PAP, adaptive servoventilation (ASV) devices provide varying amounts of pressure when breathing drops below a certain level as measured over a few minutes. Their clinical role is still evolving at this time due to concerns about effectiveness and possible increase in ultimate cardiovascular mortality in some patients with heart failure. Nasal oxygen administration may decrease central apneas in some patients, though it may increase the number of obstructive events. Some doctors use the drug acetazolamide to lower the number of central apneas, while possibly increasing obstructive ones. Overall, treatments for central sleep apnea are less well developed than for OSA.

I LAY ME DOWN AND SLUMBER

I lay me down and slumber
And every morn revive.
Whose is the night-long breathing
That keeps a man alive?

When I was off to dreamland
And left my limbs forgot,
Who stayed at home to mind them,
And breathed when I did not?

.....

I waste my time in talking,
No heed at all takes he,
My kind and foolish comrade
That breathes all night for me.

From ***More Poems*** by A. E. Housman XIII (1936)

The Victorian/Edwardian poet A.E. Housman (1859–1936) is perhaps best known for *A Shropshire Lad*, (1896), a collection of poems capturing the feel of youth growing up in the English countryside. In this passage, he addresses another topic, and in a poetic way instinctively expresses the notion that there must be a mechanism by which the body remembers to breathe when one is asleep. As described in the text, central sleep apnea is a condition in which this mechanism fails periodically, resulting in episodes of cessation of airflow through the nose and mouth, and lack of movement of the chest wall and diaphragm.

RESTLESS LEGS SYNDROME (RLS)

Uncomfortable sensations in the legs at night

Restless legs syndrome (RLS) symptoms occur in 5–10 percent of the population, and about 3 percent are significant enough to be considered in need of treatment, making it one of the more common sleep disorders. The core symptom is an urge to move the legs, usually accompanied by an unpleasant "creepy-crawly" or tingly sensation. Key features, which help separate it from other disorders are that it occurs when at rest or lying down, is worse at night, and is relieved at least temporarily by movement such as getting up and walking. It tends to occur in middle or older age, though it can occur earlier, and is more frequent in women than men. It can appear in pregnancy, usually getting better after giving birth. It tends to run in families, and 40–90 percent of people with RLS have a close relative with the disorder. RLS is usually a long-term difficulty, often getting worse over time, though some people develop it and continue at the same moderate level indefinitely.

Diagnosis is not always easy, partly because patients often have difficulty expressing what the sensations are like, and also because RLS must be separated from a variety of other disorders that may mimic it. Among these are claudications—pains in the legs due to blocked arteries. These usually are brought on by exercise, but in more severe form can occur at rest. Unlike RLS, however they are not relieved by moving about (quite the opposite) and are not more common at night. Peripheral neuropathy (disorders of the nerves outside the central nervous system) can cause numbness and tingling feelings, but are generally not relieved by movement. Discomfort from being in an awkward position is usually relieved by a single movement, and does not necessarily occur at night. Other causes that need to be ruled out include foot tapping due to anxiety, muscle aches, akathisia (motor restlessness) as a side effect of antipsychotic drugs, and RLS symptoms brought on or made worse by some medications.

On polysomnography, most patients are found to have significant numbers of periodic limb movements, and many show evidence of excessive body movements and decreased sleep efficiency and total sleep. In addition to being an unpleasant experience, it can have a negative impact on quality of life, and is associated with a higher risk of cardiac disease, as well as depression and anxiety. An MRI study has suggested that people with RLS may have a higher incidence of silent small vessel disease, which is considered a risk factor for stroke.

The cause of RLS has not been fully elucidated. It has long been associated with iron deficiency anemia, and more recent studies have indicated that there may be iron deficiency in the brain, even though the amounts of iron in the circulating blood are not necessarily low. In the brain the deficiency seems to center on a structure known as the substantia nigra (see pages 34–5), a dopaminergic center, and it seems likely that this leads to an alteration in dopamine function. RLS

FACTORS INCREASING THE LIKELIHOOD IN ADULTS OF HAVING RESTLESS LEGS SYNDROME

Increasing age	Smoking
Lower income	Pregnancy
Minimal exercise	Iron deficiency
Higher body mass index	Renal impairment

WHEN LEGS WON'T REST
RLS involves an uncomfortable feeling of the legs (and sometimes upper limbs) that has been described as "creepy-crawly," tingly, or "pins and needles" and is relieved by movement. It occurs at rest, primarily at night, and can range in severity from mild to extremely disturbing. It is one of the most common sleep disorders, occurring to various degrees in 5–10 percent of the population.

may also be associated with other medical conditions ("secondary RLS"). In addition to anemia, these include uremia and renal failure, Crohn's disease, and celiac disease. Some medications may possibly make RLS worse, including antihistamines and antipsychotics. There have been reports that this may be true of antidepressants as well, though this is now less clear.

Many doctors treat RLS with L-DOPA or dopamine receptor agonists (stimulators) such as pergolide, ropinirole, or rotigotine. The latter can be administered by a slow-releasing skin patch. Sleepiness sometimes occurs with these type of agents. Interesting, a small percentage of patients develop increased urges for sexuality or gambling, which is also known to happen when Parkinson's disease patients take these medications. Other drugs, generally less effective, include some anticonvulsants, benzodiazepines ("Valium-like" drugs), and opioids. In addition to medication treatment, it is important to maintain good sleep hygiene, as many people develop the complication of conditioned insomnia described in Chapter Eight.

PERIODIC LIMB MOVEMENT DISORDER

Jerking movements of the limbs during sleep

Many people with sleep disorders are restless during sleep, tossing and turning throughout the night. In contrast, some individuals periodically have very specific, discrete movements of the limbs, in a condition known as periodic limb movement disorder (PLMD). Most commonly these involve a very brief flexion of the hip, knee, and ankle, lasting 0.5–10 seconds, occurring about every 20–40 seconds. Such movements appear in about 12 percent of people with insomnia and 3 percent of those with excessive sleepiness. They are more common in people over 40, but can appear in young adults and even children. In considering the possibility that a person has this disorder, it can be helpful to ask the patient's bed partner whether (s)he is aware of such movements.

Because these movements are often accompanied by signs of arousal on the EEG (alpha waves or K-complexes, as described in Chapter One), they have traditionally been thought to disrupt sleep and be a cause of clinical sleep disturbance. PLMD is diagnosed by the polysomnogram, using electromyographic

MEASURING LIMB MOVEMENTS
When undergoing a polysomnogram, leg movements are measured by placing pairs of electrodes on the skin over the anterior tibialis muscles near the shin.

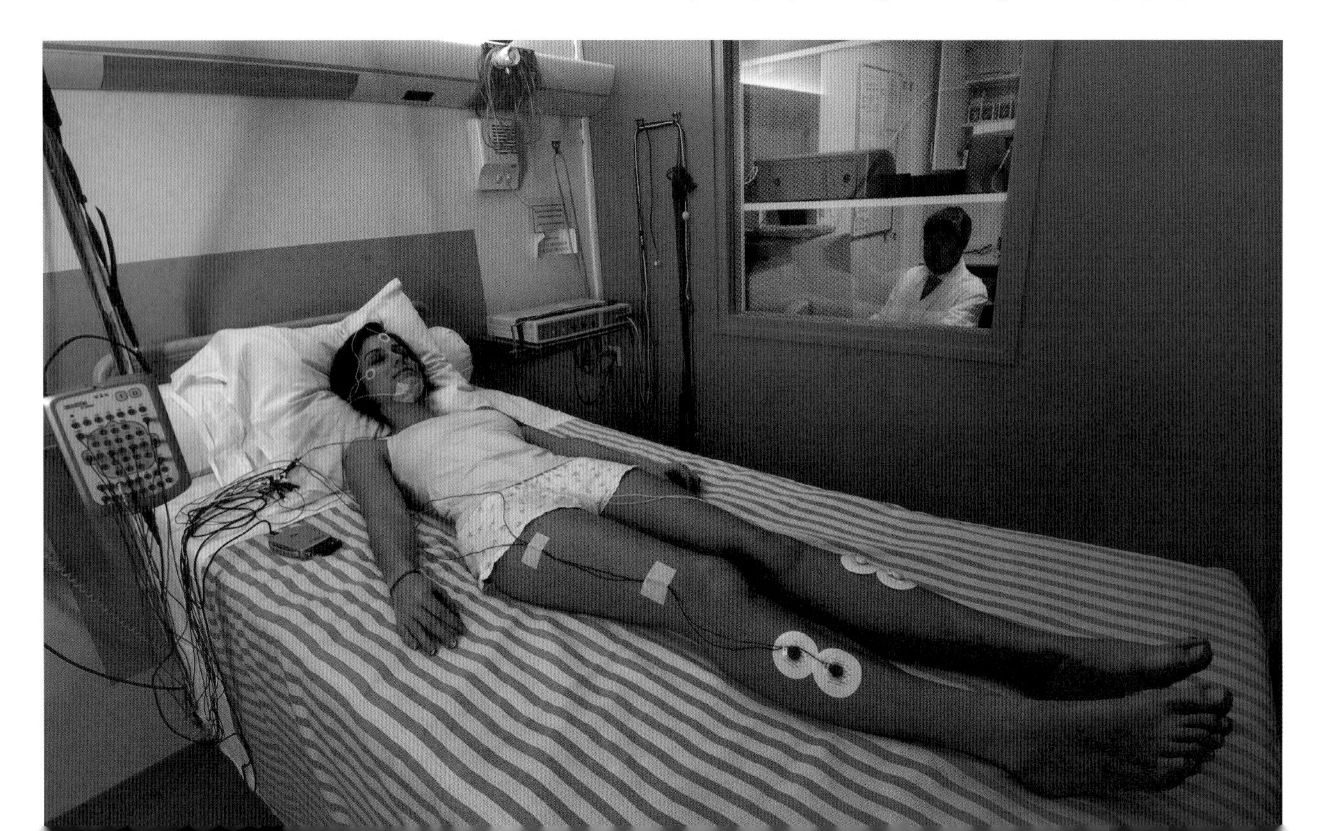

POLYSOMNOGRAM OF PERIODIC LEG MOVEMENTS

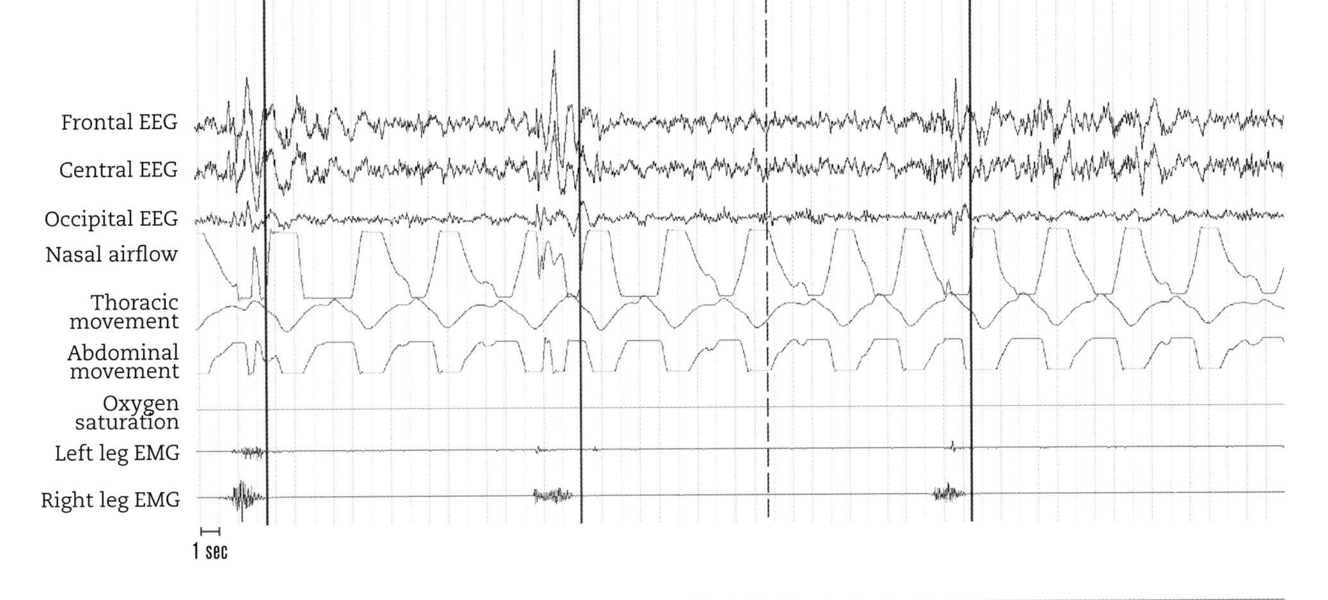

PERIODIC LEG MOVEMENTS
In PLM disorder, there are brief (0.5–10 seconds) movements of the legs. In this particular person, the movements are more on the right side than the left. These movements are periodic, occurring every 20–40 seconds. A diagnosis requires movements to occur at a rate of at least 5 per hour in children and 15 per hour in adults.

electrodes placed over the anterior tibialis muscle (on the shin of the lower leg). These movements may be found in other disorders including narcolepsy, REM behavior disorder, and particularly in restless legs syndrome in which it may be present in 80–90 percent of cases. They may also be related to some medications, including tricyclic antidepressants and selective serotonin uptake inhibitor antidepressants, though studies about this have been variable. When they occur in patients with sleep disturbance in the absence of these other disorders or medications, a diagnosis of PLMD is considered. The diagnosis requires finding a rate of 5 per hour of sleep in children and 15 per hour in adults. Treatment is typically with the benzodiazepine ("Valium-like") drug clonazepam, which seems to reduce the number of arousals associated with the movements. A common side effect of clonazepam, however, can be sedation.

The thinking about PLMD is evolving. Originally it was described as "nocturnal myoclonus," (involuntarily twitching of muscles) which may have been a poor choice as it seemed to suggest that this was associated with a form of epilepsy. Later it came to be recognized as the occurrence of movements which caused arousals, as we presented it above. In more recent years, a new approach emphasizes a failure during sleep of descending signals from the brain that decrease spinal excitability. Normally during sleep there is increased inhibition of spinal leg-flexing reflexes. It is thought that during the night periodic activation signals, coinciding with what are known as EEG cyclic alternating patterns (cycling between deeper and lighter sleep) occur, and in PLMD are inadequately inhibited, resulting in the leg movements. In this view, the leg movements are not the cause, but rather the result, of arousals. With this approach, periodic leg movements are seen less as comprising a disorder per se, but rather may represent a marker of risk of other disorders, particularly restless legs syndrome.

EXCESSIVE SLEEPINESS DUE TO NERVOUS SYSTEM DISORDERS

When nervous system malfunction makes us too sleepy

Although insomnia is the most common sleep complaint in the general population, excessive sleepiness is the most likely reason to bring a person to a sleep disorders clinic. Very often, the history of being too sleepy is accompanied by symptoms suggestive of obstructive sleep apnea, such as snoring, or witnessed periods of breathing cessation accompanied by high blood pressure, in which case it is appropriate to do laboratory studies of respiration and blood oxygenation in sleep. In the absence of these respiratory findings, there are a wide range of other types of disorders of excessive sleepiness. These include narcolepsy with cataplexy (a sudden loss of muscle tone in association with emotion), which will be the focus of this section, as well as narcolepsy without cataplexy, central nervous system hypersomnolence disorder, sleepiness due to lifestyle sleep deprivation, and medical conditions or drugs. Here we have chosen to explore narcolepsy with cataplexy because it is a good example of how sleep science has begun to clarify its underpinnings both at phenomenological (the "REM onset sleep") and physiological (hypocretin deficiency) levels. There is also an interesting animal model—dogs who have an analogous condition—which facilitate its study. Because of its intriguing qualities, characters with symptoms suggestive of narcolepsy have been portrayed in classical works of fiction including Edgar Allen Poe's *A Premature Burial*, George Eliot's *Silas Marner*, and Herman Melville's *Moby Dick*.

Narcolepsy–cataplexy

Narcolepsy–cataplexy, which occurs in 0.02–0.04 percent of the population, is an illness that is characterized by excessive daytime sleepiness and cataplexy, accompanied by one or more secondary symptoms including sleep paralysis, hypnogogic hallucinations, and disrupted nocturnal sleep. These symptoms frequently do not begin at the same time, and an individual person with narcolepsy does not necessarily have to have all of them. A typical history might be of an adolescent or young adult who complains of excessive sleepiness which cannot be explained by lifestyle or drugs, whose inappropriate episodes of sleep puzzle all concerned. After a few years, cataplexy or some of the secondary symptoms then appear as well, at which time the diagnosis is made. A description of the individual symptoms follows.

OTHER FEATURES ASSOCIATED WITH NARCOLEPSY

In addition to the symptoms such as excessive sleepiness that are described in the text, people with narcolepsy are at higher risk for several associated conditions. These include:

- Depressive symptoms
- Increased body mass index and waist size
- Decreased quality of life
- REM behavior disorder (see page 133)
- Sleep-related eating disorder

Very significant sleepiness

People with narcolepsy are remarkably sleepy. They may fall asleep in the middle of a conversation, when heavily engaged in some activity, even during sex. Indeed the key to a history of this kind of pathological sleepiness is to inquire whether a person has fallen asleep at times when it would be highly unusual to do so. Characteristically, when they do fall asleep, they often awaken from a short nap feeling at least temporarily refreshed.

Cataplexy

Cataplexy, as mentioned earlier, is a sudden loss of muscle tone in the major weight-bearing muscles, as well as head and neck, which occurs in association with emotion. It may involve minor weakness leading a person to need to sit down, but often is much more dramatic, causing a fall to the floor. The key feature is the association with emotion, which may be laughter or anger or, for instance, seeing an exciting movie. Some researchers believe that it is an extreme case of a phenomenon we all feel, captured by such phrases as "We had them rolling in the aisles" or "My jaw dropped open when he said that." The episodes are very brief, usually from 30 seconds to a few minutes. After longer cataplectic episodes, patients may describe having experienced visual hallucinations.

SLEEPING ON THE JOB

There are of course a wide variety of reasons for being too sleepy, including sleep deprivation, circadian rhythm disorders, medications, and toxic/metabolic conditions. One important one is the group of disorders of the nervous system leading to excessive daytime sleepiness, including narcolepsy and idiopathic hypersomnia.

Sleep paralysis

In sleep paralysis, which usually occurs in the transition between sleeping and wakefulness, a patient feels conscious but unable to move for periods of a few seconds up to perhaps 20 minutes. This is often accompanied by a feeling of anxiety or panic. In contrast to cataplexy, a person can easily be aroused from an episode of sleep paralysis. It is often accompanied by visual or auditory hallucinations, sometimes a sensation of a visualized figure, a person or creature, sitting on one's chest. Sleep paralysis, though very much part of narcolepsy when seen in the context of extreme daytime sleepiness, can also occur in normal individuals.

Hypnogogic hallucinations

Hypnogogic hallucinations are vivid, dream-like visual experiences that occur at the transitions between wakefulness and sleep. They are in contrast to the more thought-like reverie that often accompanies falling off to sleep. As mentioned before, they often occur in the context of cataplectic episodes.

Disturbed night-time sleep

Paradoxically, these same individuals who are so very sleepy in the daytime have a difficult time at night, with frequent awakenings and tossing and turning.

General features of narcolepsy–cataplexy

Narcolepsy–cataplexy affects both sexes about equally, most typically appearing between ages 10 and 20, although about one-quarter of cases begin after age of 20. It is an illness that can have profound effects on a person's life. About three-quarters of people with narcolepsy feel that their sleepiness presents major problems at work, and about two-thirds describe falling asleep when driving or having near-misses due to sleepiness. Narcolepsy is also often accompanied by other conditions including obesity, psychiatric disorders, and reported cognitive disturbances. Childhood onset narcolepsy–cataplexy is associated with increased likelihood of being overweight and developing premature puberty. One possible mechanism for this may involve decreased activity of the orexin/hypocretin system (see pages 36–7), which may influence release of sex hormones. Depressed mood is very common, although the relationship of narcolepsy and true major depressive illness remains a matter of research by means of epidemiological studies. Some patients complain of difficulties with memory, although it is not clear whether this is inherent to the illness, or whether it is a result of missing the original input of material during micro-sleep episodes.

Narcolepsy is generally a lifetime disorder, though it may ameliorate somewhat in older age. There appears to be some genetic component to it, insofar as about 0.9–2.3 percent of close relatives manifest the illness, and a much higher number report problems with excessive sleepiness. Insofar as only about 35 percent of identical twins both have narcolepsy, some other non-genetic factors such as the environment and overall health may play a role in its genesis. It has been found to be associated with markers on human blood cells involved in immune function, and about 90 percent of patients are positive for what is known as an HLA antigen DQB1*0602. It is not diagnostic, though, as this marker also appears in significant numbers in the general populations of many different ethnic groups, from 12 percent in Japanese to 38 percent in African-Americans. The association of narcolepsy with this white blood cell marker has led some researchers to suggest that narcolepsy may be at least in part an auto-immune disorder. Supporting this notion has been the observation of a relation to infections including group A streptococcus and influenza A, as well as some flu vaccines used in Europe, which might potentially be precipitants of this process. As mentioned earlier, narcolepsy with cataplexy is associated with a decrease in the wake-promoting peptide orexin/hypocretin in the hypothalamus. Indeed, low levels of the orexin/hypocretin peptides in the cerebrospinal fluid can be observed, and are used as a confirmatory diagnostic step. Levels appear to be normal in idiopathic hypersomnia.

Idiopathic hypersomnia

As its name indicates, this is a disorder of excessive sleepiness without known cause. These sleepy people do not have cataplexy. In contrast to narcoleptic patients, who may take short naps and at least temporarily feel refreshed, they may have long

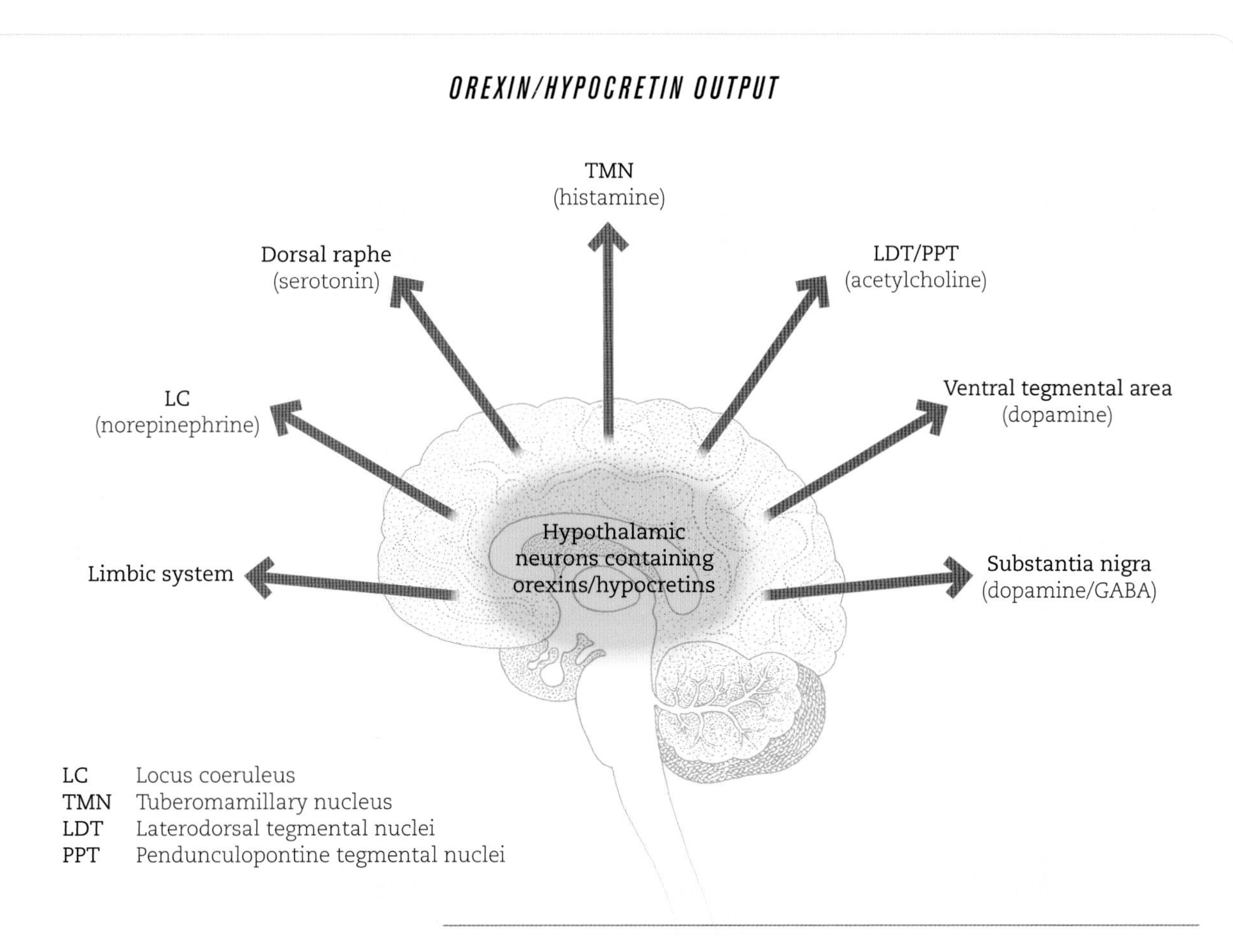

OUTPUTS OF THE OREXIN/HYPOCRETIN SYSTEM

Orexin/hypocretin-containing neurons in the lateral hypothalamus project to other areas of the hypothalamus, the brain stem (including the locus coeruleus and dorsal raphe), the limbic system and thalamus, and more widespread areas of the brain. They tend to increase wakefulness and prevent inappropriate appearance of REM sleep, as well as having other effects including increased appetite. Orexin/hypocretin pathways increase activity of norepinephrine and serotonin, which among many other activities inhibit the descending pathways that decrease muscle tone in REM. When there is an orexin/hypocretin deficiency, as in narcolepsy, decreased monoaminergic activity allows cataplexy to appear when awake.

periods of daytime sleep, from which they often wake up irritable and unrefreshed. Although narcolepsy is generally a lifelong illness, some people with idiopathic hypersomnia spontaneously get better. Narcolepsy is diagnosed by the appearance of "REM onset sleep" on the polysomnogram; this phenomenon does not occur in idiopathic hypersomnia, though they do show sleepiness as manifested by an average sleep latency of less than eight minutes. There are of course many other conditions that can cause excessive sleepiness, including insufficient sleep, sleepiness due to medicines or medical conditions, and menstrual-related hypersomnia. In Kleine–Levin syndrome, adolescents, particularly boys, have long periods of sleepiness with a fluctuating course, often accompanied by binge eating and very increased sexual drive and depressed mood.

DIAGNOSING AND TREATING EXCESSIVE SLEEPINESS

Assessing narcolepsy and idiopathic hypersomnia

In Chapter Three we talked about vigilance and performance tests in assessing sleepiness. There are also scales that assess the subjective degree of sleepiness, notably one known as the Epworth Sleepiness Scale. Physiological sleepiness can be measured by use of the multiple sleep latency test (MSLT). This is done by offering the patient four or five 20-minute nap opportunities across the daytime. The sleep latency (time from when the lights are turned out until sleep onset) is determined, and the average of the sleep latencies is recorded. People with normal wakefulness will usually have average sleep latencies of 10 minutes or longer. Sleepiness is considered to be the case when there is an average sleep latency of less than 8 minutes; and 5 minutes or less is considered pathological. People with narcolepsy will typically have mean sleep latencies of 8 minutes or less, and in addition will have what are known as REM onset sleep episodes. As was discussed in Chapter One, typically in normal adults REM sleep does not appear until 90 or more minutes after sleep onset. In people with narcolepsy, REM may appear within 15 minutes during these brief naps on the MSLT, and having two of them is considered crucial to the diagnosis. As with so many things in medicine, the diagnosis includes not only this laboratory finding, but also the overall clinical picture. This is important, because REM onset sleeps are also seen in sleep apnea patents and sometimes in normal sleepers. As mentioned above, people with idiopathic hypersomnia will have the excessive sleepiness as seen by a short average sleep latency, but not the REM onset sleep episodes. There are also variations on the MSLT including the maintenance of wakefulness test (MWT), which is conducted in the same way except that the subject is instructed to try to stay awake. Some medical researchers feel that the MWT is more predictive of the consequences of sleepiness on daytime life, while others feel that this is not clear.

Treatment of narcolepsy and idiopathic hypersomnia

Narcolepsy and idiopathic hypersomnia can be treated both with medicines and also by behavioral interventions. In most cases both approaches need to be used at the same time.

Non-medication approaches

Several behavioral steps need to be considered in people with narcolepsy, including maintaining good sleep hygiene (see pages 154–5). Daytime naps can be very beneficial and, as mentioned earlier, it is characteristic of narcolepsy that patients awaken feeling refreshed and alert, at least temporarily. Sometimes people with narcolepsy should address this need for a brief daily nap with their employers. Overall, they have often not achieved as much as they would like in their professional and personal lives, and have poor self-esteem. Many—more than in the general population—have symptoms of depression that can be treated either by psychotherapy or pharmacotherapy. As it happens, some antidepressants are used by doctors to treat cataplexy, so they are sometimes used both for this purpose and for depressed mood.

Medication treatment

Doctors use a variety of medications to treat different aspects of narcolepsy. For sleepiness the original drugs have been stimulants such as methylphenidate or dextroamphetamine. Their

ASSESSING SLEEPINESS

On the Epworth Sleepiness Scale, subjects are asked to rate the likelihood of their dozing off or falling asleep in eight situations—including sitting and reading, watching TV, talking to someone, and sitting in a car while stopped for a few minutes in traffic. They are asked to rate the likelihood of dozing on a scale of 0–3 in which 0 represents no chance of dozing and 3 a high chance of dozing. The numbers are added up, with the highest possible score being 24. In general, the normal range is 0–10, 11–12 indicates mild excessive sleepiness, 13–15 moderate excessive sleepiness, and 16–24 severe sleepiness. People with narcolepsy generally score in the moderate or severe sleepiness range.

effectiveness is limited by some drawbacks including their dependence-producing properties (they are controlled substances), liability for abuse, and side effects including possible high blood pressure and cardiac disturbances. Modafinil and the longer-acting related compound armodafinil are often effective, and have a generally more benign side effect profile, but also include rare but potentially medically serious rashes and allergic reactions, as well as psychiatric or cardiac effects. For treatment of cataplexy, antidepressants such as venlafaxine and fluoxetine are often used. The only drug currently available that addresses both sleepiness and cataplexy is sodium oxybate. It may also aid in the disturbed night-time sleep as well. Limitations are that it has a history of being a drug of abuse, is a controlled substance, and is available only through specially licensed pharmacies. Patients also typically need to be awakened once during the night to take an additional dose. It is a nervous system depressant and should not be used in patients taking other sedatives or alcohol, and may depress respiration, particularly in people with other pulmonary disorders or sleep apnea. As with all the medications mentioned above, it also has the potential for affecting judgment and alertness when driving.

Pharmacologic treatments for idiopathic hypersomnia have generally been less effective than for narcolepsy. Stimulants and some antidepressants have been used. Many doctors feel that modafinil or armodafinil are the treatments of choice, though they are not specifically approved for this purpose.

PARASOMNIAS

Undesired behaviors or experiences during sleep

Parasomnias are considered to be disorders of arousal, partial arousal, or transitions between sleep stages. Some parasomnias are associated with NREM sleep, while others begin in REM sleep, and we will consider each in turn.

Disorders of arousal from NREM sleep

We have so far described the sleep stages, and sleep itself, as fairly distinct states, but in some clinical conditions the lines may become blurred. This has sometimes been phrased by saying that sleep and waking are not always mutually exclusive conditions. In disorders of arousal, some behaviors typical of waking (ambulating, talking, eating) may occur out of sleep, and some features of sleep (diminished consciousness and executive function) may continue into the transition into waking. The most common of these is confusional arousals, which can be seen in children but are more common in late adolescence and young adulthood. These people wake up disoriented to the time and where they are, are sluggish, and do not make much sense when answering questions. On rare occasions this confusion can lead to violence. As with most of the disorders of arousal, there is little or no memory of the event later. In the majority, other sleep disorders such as sleep apnea are present, and a significant minority have psychiatric disorders including anxiety and bipolar disorder. The presence of sleep apnea points to a feature common to the disorders of arousal: they are often associated with conditions that cause disturbed sleep, whether due to sleep disorders or environmental conditions such as excessive noise. Disorders of arousal are also more likely to appear when combined with processes that lead to deeper sleep, which can include prior sleep deprivation, or use of a sedative medication, typically the newer non-benzodiazepine GABA receptor agonist hypnotics. These medicines, which are discussed in detail in Chapter Eight, are widely used to aid sleep, and generally are more benign than older types of sleeping pills, but of course have some limitations themselves. There is also now the suggestion that people with all NREM parasomnias may have some common genetic features, as they share a high prevalence of a particular genotype, a gene sequence on white blood cells known as HLA DQB 1*05:01.

There are no specific treatments for confusional arousals, aside from safety precautions such as covering windows, and getting adequate amounts of sleep. There is some disagreement about whether alcohol may be a precipitant, but many clinicians stress the need to avoid alcohol and sedative medications.

Sleepwalking

Sleepwalking is usually thought of as a childhood disorder, but actually it occurs occasionally in about 5–6 percent of adults. It usually happens in the first third of the night. Typically the sleepwalker returns to bed, with no memory of the event the next day. As with all the arousal parasomnias, the sleepwalker often appears confused and disoriented if awakened by others during the event. They most commonly have some responsiveness to the environment, not walking into walls, for instance, though of course for safety reasons it is important to remove possible hazards from the sleep location. There is a strong genetic component as well, and it can be provoked by prior sleep deprivation. Although it can occur during N2 (stage 2) sleep, it typically occurs in N3 (slow-wave sleep), and during the event itself N3 sleep may be seen on the EEG. One diagnostic clue on the sleep study is that there are often frequent brief arousals seen during N3 sleep. There are no generally accepted effective treatments. Avoiding precipitants such as sleep deprivation can be

THINGS THAT GO BUMP IN THE NIGHT
The Somnambulist (1871), by the English painter John Everett Millais (1829–96), is thought to be taken from *La Sonnambula*, an opera by Bellini about a young woman whose sleepwalking presents romantic difficulties. Here the sleepwalker holds a candelabra which has gone out, leaving the scene even darker as she walks along the edge of a cliff. Although sleepwalking is most common in children, it does occur in adults as well. It most typically occurs in stage N3 sleep, and is made more likely under conditions of sleep restriction. One aspect of caring for a person sleepwalking is to reduce hazards in the environment such as open windows or staircases. In most cases of NREM sleepwalking, however, the person returns to bed, with little or no memory of the event in the morning.

important. There may be some association with the use of lithium, and possibly the newer non-benzodiazepine GABA agonist sleeping pills (see page 160). Management includes making the environment as safe as possible; sometimes hypnosis has been used, and some clinicians prescribe benzodiazepine sedatives such as clonazepam. Again, a possible side effect can be sedation.

Night terrors

Night terrors typically occur in children as well as some adolescents. They are characterized by the child letting out a loud scream and displaying intense fear. Signs of autonomic arousal such as rapid heart rate and respiration are prevalent. Usually the child is not aroused easily, is not consolable, does not report dream-like content, and ultimately returns to sleep with no memory of the event the next morning (though often producing a significant memory in the parents). These events are usually in the first half of the night, and typically occur out of N3 sleep, which is often fragmented on the sleep study. Sleep apnea and restless legs syndrome are often found as well. Again, there is a strong genetic component.

Management of sleep terrors includes ensuring that the person is not sleep deprived, and addressing sources of anxiety or stress. Naps can be useful, and some find that arranging for scheduled awakenings during the night is helpful. There are no specific medicine treatments, though some clinicians give benzodiazepine hypnotics, while others, with relatively little systematic evidence, try the selective serotonin reuptake inhibitor (SSRI) medicine paroxetine. Sleep terrors also need to be distinguished from nightmares, which are described next.

DISTURBING DREAMS
Although many people have disturbing dreams from time to time, in nightmare disorder they are frequent and intense. Nightmares are a REM sleep phenomenon. In contrast to NREM sleep terrors, the person generally awakens fully, is lucid and remembers the contents of the dream.

REM-related parasomnias

In contrast to sleepwalking and night terrors, which tend to occur out of N3 sleep, some other parasomnias usually appear in association with REM sleep.

Nightmare disorder

A person with nightmares awakens from sleep with a vivid memory of a very disturbing dream, which can involve fear but also other emotions such as anger or disgust. Although most adults describe having nightmares once in a while, the sufferer from nightmare disorders has these frequently. A number of features separate nightmares from sleep terrors. They are more likely to occur in the second half of the night. In contrast to sleep terrors, the dream is usually remembered. Generally the person is lucid, not confused, and has only mild symptoms of autonomic stimulation such as fast heartbeat and respiration. Nightmares are a REM sleep phenomenon, which fits with their predominant appearance in the second half of the night, when REM is more plentiful. People who are sensitive, or who are having difficult long-term relationships may be particularly likely to experience nightmares. For those requiring treatment, imagery-oriented or cognitive psychotherapies are often used. Some doctors prescribe the older tricyclic antidepressants such as amitriptyline in order to reduce nightmares, and a variety of other medicines including the blood pressure medicine prazosin have been tried, though they are not specifically approved for this purpose.

REM behavior disorder

In Chapter Two in the sections "Sleep and dreaming in animals" and "REM sleep mechanisms" (pages 50–1) we discussed the physiologic processes by which descending signals from the brain stem inhibit muscle activity during REM sleep, and described how lesions of areas in the pons of the brain stem can disrupt this system, resulting in animals behaving as if they were acting out their dreams. In humans, there is a syndrome known as REM behavior disorder (RBD) in which sufferers get up during the night and engage in motor behavior, involving recognizable activities and sometimes violence to others. The cause appears to be a failure of the normal mechanism inhibiting muscle activity during REM. RBD can be either idiopathic (of unknown origin) or result from withdrawal from alcohol or sleeping pills, or be a rare side effect of most types of antidepressant medication. It typically occurs in older people, and many go on to develop Parkinsonism or degenerative brain disorders in the next few years after diagnosis. It is sometimes seen in people with narcolepsy–cataplexy. There are often reports of family members acting out dreams. RBD can be a very serious problem, particularly in view of the predilection for violence. Although it is difficult to actually capture an event occurring out of REM during a sleep study (though it has been done), a suggestion of RBD is found in the observation that they often have less atonia (relaxation of muscles) or more brief movements on the EMG during REM sleep. Management of RBD involves making the environment as safe as possible. Many doctors prescribe the benzodiazepine ("Valium-like drug") clonazepam, or melatonin (see pages 78–9).

Recurrent sleep paralysis

Earlier in this chapter we described the symptom of cataplexy as part of the narcolepsy–cataplexy disorder. It turns out that some people, typically young adults, with normal daytime wakefulness and no evidence of narcolepsy have recurrent episodes of sleep paralysis. The cause is unknown and there is not necessarily an association with psychiatric disorders. If treatment is needed some doctors give the older tricyclic antidepressants or the selective serotonin reuptake inhibitor medications.

NIGHT MARES?

The word "nightmare" is not related to horses, but rather comes from the Old English term *maere* as well as similar words in old Dutch and German, referring to a malign female spirit or incubus who would afflict sleepers with a sense of difficulty breathing. Over the years its meaning evolved to mean bad dreams sent from the incubus, and ultimately to just scary dreams in general.

MISCELLANEOUS DISORDERS

Rhythmic movements, seizures and headaches

In this section we will briefly describe several types of sleep disorder that do not fit into the other categories described earlier in this chapter; these include rhythmic movement disorder, sleep-related seizures, and sleep-related headaches.

Rhythmic movement disorder

This disorder, typically of infants in the first year of life, involves repetitive stereotyped motor behavior, such as head banging or body rocking. Episodes usually last just a few minutes, but in some cases can go on for several hours, and usually occur when the infant is alone in bed or crib. These movements usually occur in NREM sleep, although they can be in both NREM and REM, or REM alone. Aside from padding the walls of the crib, treatment is not usually needed. Behavioral treatments have had limited success.

Sleep-related seizures

About one-fifth of patients with seizures have them only during sleep, and an additional one-third have them during both sleep and waking. Seizures that occur during sleep only tend to happen at two peak times, at about 10:00–11:00 PM and 4:00–5:00 AM. Different types of seizure have their own likelihood of occurring during different sleep stages, but in general they are more likely in NREM, often in N2 (stage 2). The location of the type of seizure in the brain also is related to when it occurs; for instance, epileptic events starting in the frontal lobe of the cortex tend to occur in sleep, while those in the temporal lobe tend to start in wakefulness. Epilepsy may influence sleep in several different ways, including affecting sleep due to the seizure itself or due to seizure-like patterns of the EEG without actual seizures, or as a result of anticonvulsant medications. Often in generalized seizures there is reduced total sleep and REM sleep. Indeed, even on nights without seizures there is often reduced sleep efficiency and an increased number of awakenings. One particular type of epilepsy, nocturnal frontal lobe epilepsy, may involve multiple arousals or abnormal movements in bed, but some may get up and engage in walking or running. This needs to be distinguished from NREM sleepwalking. Ultimately the EEG and a history of daytime seizures differentiate the two, but there are also some differences in clinical appearance. Sleep-related seizure behaviors can occur throughout the night instead of primarily during the first part; there is a higher likelihood of accidental injury, and the person involved is less likely to return to bed.

Doctors usually treat seizures with antiepileptic medication, which can cause daytime sleepiness. Obstructive sleep apnea often also occurs in these persons, and treating the sleep apnea may sometimes have beneficial effects on the seizure disorder.

Sleep-related headaches

Headaches from a variety of causes can result in disturbed sleep and daytime sleepiness. Some people, usually middle aged or elderly, have "alarm clock headaches" (hypnic headaches) which awaken them at about the same time each night. Cluster headaches involve pain associated with one eye, with tearing and nasal congestion. These may appear out of NREM or REM sleep, and sometimes also happen at about the same time each night. Migraines can be associated with sleep, and tend to occur out of REM. Some sleep disorders such as obstructive sleep apnea can result in headaches, usually upon awakening, and the treatment for these usually focuses on the sleep apnea itself.

NOCTURNAL FRONTAL LOBE EPILEPSY

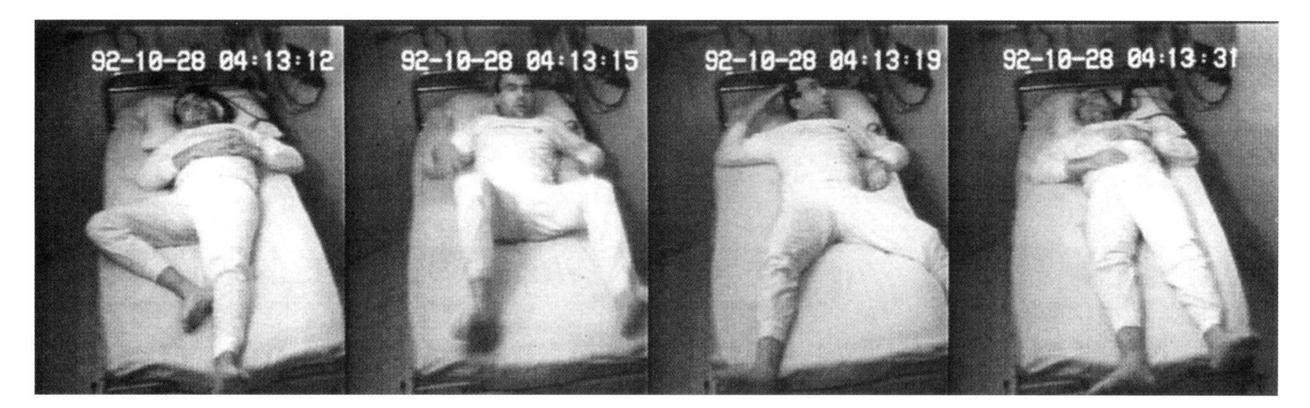

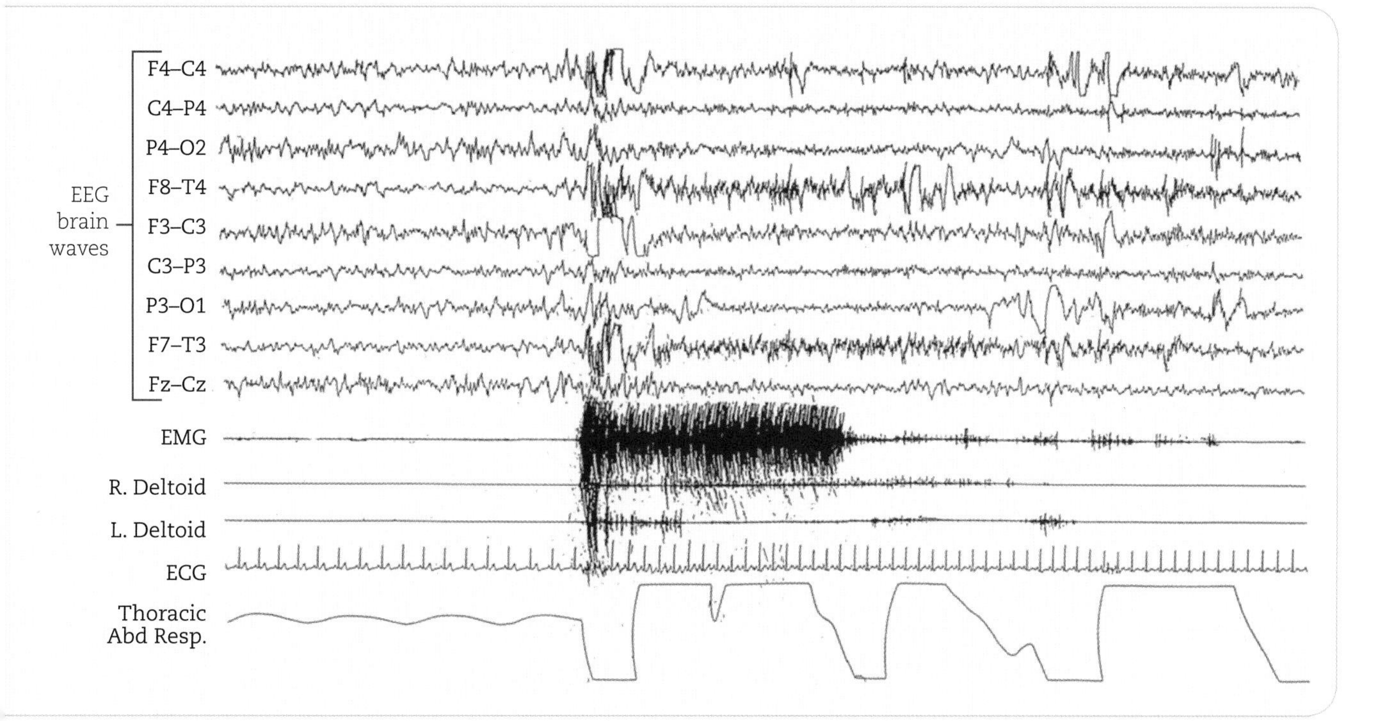

Key

F Frontal lobe
T Temporal lobe
C Central lobe
P Parietal lobe
O Occipital lobe

Each EEG channel represents the difference in electrical activity between two locations on the scalp (bipolar recording).

ECG Electrocardiogram

SEIZURES DURING SLEEP

This illustrates a type of seizure known as "paroxysmal arousals," which occur out of NREM sleep and usually last less than 20 seconds. In this one, the person begins to sit up, opens his eyes, appears to be frightened, and returns to sleep. Note that a K-complex appears immediately before the epileptic activity seen in the EEG channels. Republished with permission of Oxford University Press.

SLEEP IN PSYCHIATRIC ILLNESS

Sleep in depression and anxiety disorders

The American sleep researcher Frederick Snyder (1938–2003) summed up a complex relationship by saying that troubled minds have troubled sleep, and troubled sleep causes troubled minds. It is certainly true that insomnia is very common in people with psychiatric illness. Perhaps the most frequent is in depression, in which 75 percent or more of individuals describe difficulty sleeping. In panic disorder this is about 60 percent, about 50 percent in schizophrenics, and at least 25 percent in alcohol abuse. Conversely, insomnia appears to be a risk factor for later developing depression. In a one-year follow-up study of people with insomnia, at the time of initial interview, their likelihood of having a diagnosable psychiatric disorder was 1.6 times that of the general population. A year later, about 10 percent still had difficulty sleeping. Among these individuals, the likelihood of having a psychiatric illness, most commonly depression, was almost forty times higher.

Among people with depression, continued insomnia is also a marker for risk of later suicide. In a study of about 1,000 patients with major depression, the factors in their history that predicted suicide in the long term were the ones we usually think of: reported suicidal thoughts and feelings of hopelessness, as well as a history of past suicide attempts. In contrast, the factors that best predicted suicide during the next year included insomnia, anxiety, panic attacks, and alcohol abuse. In the next section we will describe the sleep of people with depression. Later we will come back to the question as to whether insomnia is just a consequence of being depressed, or whether it might be involved in the process of becoming depressed.

CONDITIONS IN WHICH A SHORT REM LATENCY IS SOMETIMES FOUND

- Obsessive–compulsive disorder
- Anorexia nervosa
- People undergoing divorce
- Narcolepsy
- Normal people living in conditions with no time cues (see page 85)
- People withdrawing from certain drugs of abuse
- Schizophrenia
- Borderline personality disorder

Sleep in depression

One way to summarize sleep in depression is that it is short, shallow, and fragmented, with REM that occurs early, and intense eye movements. This is particularly true in unipolar depression (that is, people with depression but who do not have manic episodes), older, and more agitated people. (The depression of people with bipolar illness is a little different and more likely to be accompanied by excessive sleepiness.) It is helpful to look at these qualities of sleep in people with major depressive disorder.

Short, shallow, and fragmented

Typically total sleep time is reduced, with many awakenings during the night. The classical clinical history is that depressed people awaken early in the morning and are then unable to go back to sleep. The sleep they do get is shallow, in the sense that there is reduced slow-wave sleep.

REM phenomena

In patients with depression, the time from sleep onset until the start of the first REM period ("REM latency") tends to be shorter. In non-depressed people the first REM period of the night is relatively brief, and subsequent ones get progressively longer. In contrast, in depressed people, the first REM period is relatively long, so that there is not a progression in length as the night proceeds. During this first REM period, there is

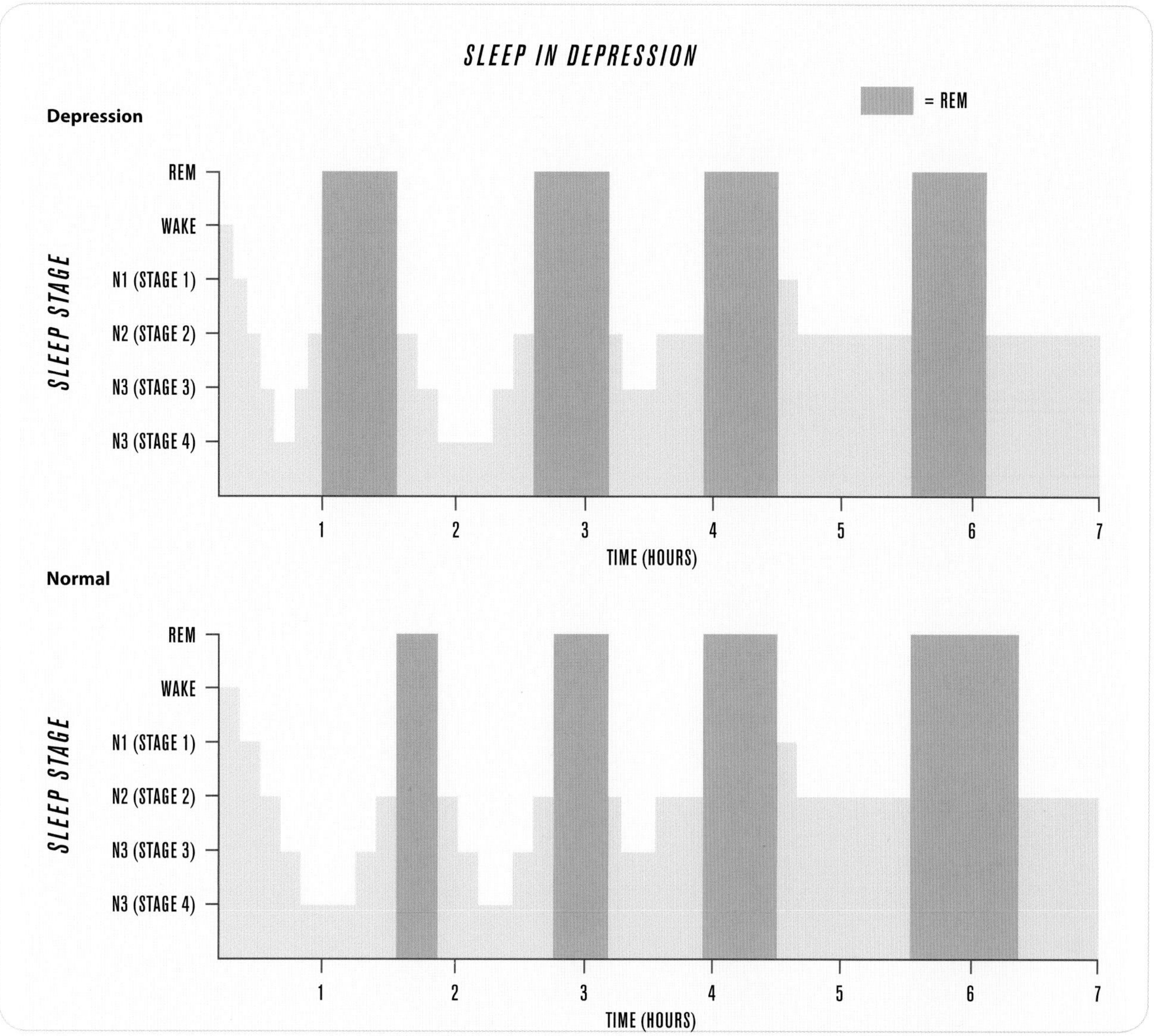

SLEEP IN DEPRESSION

an increase in eye movements per minute of REM (the "REM density"). Initially, it was thought that the short REM latency might be a biological marker for depression, something psychiatrists look for in every illness as a way of understanding its physiology. Subsequent research, however, found that short REM latencies occur in a number of conditions (see box opposite). Thus, the presence of a short REM latency is very sensitive, but not very specific, for depression. Its occurrence, as well as the decreased slow-wave sleep, has been used in understanding depression in terms of models of sleep regulation.

INCREASED REM DENSITY IN DEPRESSION

Idealized sleep histogram in normal sleep and sleep of a depressed person. Note the relatively earlier onset of REM sleep in depression. In normal sleep, the first REM period is relatively short, becoming progressively longer as the night goes on. In depression the first REM period is relatively long and often contains a higher amount of eye movements, with little progression in length throughout the night. In addition to these changes in REM, sleep is typically shorter, with more awakenings during the night, diminished stage N3 (slow-wave sleep), and often early morning awakening from which the person is unable to return to sleep.

Models of sleep regulation in depression

As was described in Chapter Three, some sleep researchers in the 1960s suggested that REM sleep abnormalities, which might occur as a result of poor sleep due to some upsetting event, led to depression. In retrospect this seems less likely for several reasons, notably that typically the decreased REM is not seen until very close to the onset of mood symptoms, and more importantly purposely depriving a person of REM sleep can actually be used as a treatment for depression. We are uncertain as to what causes the sleep changes seen in depression, but one approach is to try to understand it in terms of models that have been developed of how sleep is regulated (see pages 38–9 and 80–1). Researches have led to various speculations. One example, for instance, is that in the two-process model of sleep regulation depressed people have inadequate accumulation of homeostatic pressure ('Process S') during wakefulness. Another way is to try and manipulate sleep, and see if it has any effect on the depressive symptoms.

Altering sleep as a treatment for depression

In Chapter Three, in the section entitled "Is sleep deprivation always bad?" (see pages 68–9) we described studies which have shown that various manipulations of sleep, including purposely depriving a patient of REM sleep, total sleep for one night, or

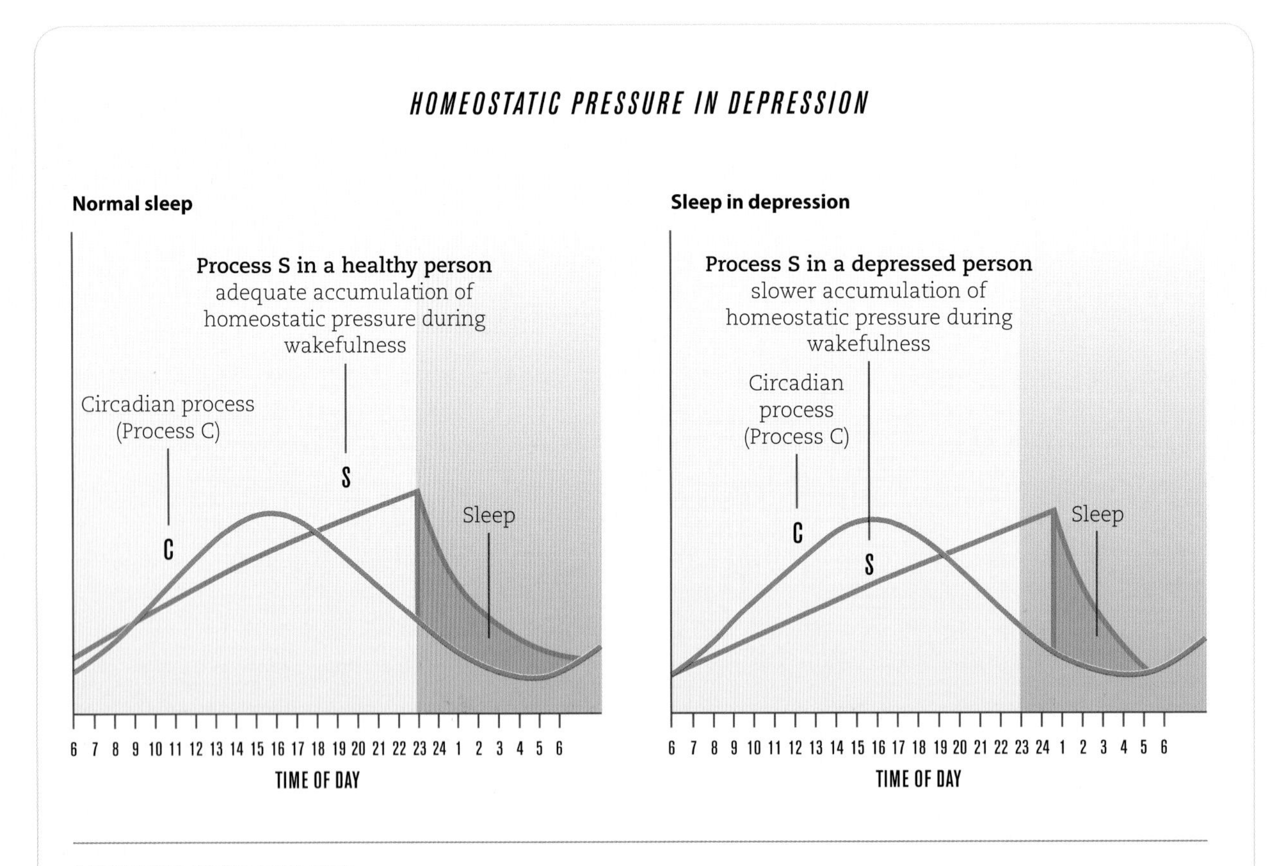

DEPRESSED SLEEP RHYTHMS

The two-process model of sleep regulation (see pages 80–1) suggests that sleep and waking phenomena result from the interaction of two activities: a homeostatic mechanism (Process S) which accumulates during waking leading to sleep, and a circadian mechanism (Process C) which includes permissive periods for sleep. The left-hand figure shows sleep beginning when Process S has accumulated during waking, and Process C is in a point permissive of sleep. In this two-process model, the delayed onset of sleep in a depressed person (right-hand figure) could be explained by a deficiency in the rise of Process S, so that it takes a longer time for it to accumulate sufficiently to help lead to sleep.

partial sleep deprivation for prolonged periods, can actually decrease depressive symptoms as effectively as antidepressants. Because of the labor-intensive nature of such procedures, they are not very practical in the real world. They make an important point though: since purposely altering sleep can actually alter the course of depression, it seems less likely that poor sleep is not just a consequence of being depressed. Instead, it makes it more likely that changes in sleep are in some manner involved in the genesis of mood disorders.

Post-traumatic stress disorder (PTSD)

Post-traumatic stress disorder (PTSD) is a condition resulting from exposure to very stressful, frightening, or distressing events. About half of people with PTSD complain of difficulty sleeping, and 50 to 70 percent suffer from nightmares. For many, night-time sleep is so difficult that they develop a fear of going to bed. Laboratory studies show decreased N3 sleep, increased light sleep (N1), and awakenings out of REM sleep. Many have increased REM density (amount of eye movements in REM sleep), and the degree of this increase is greatest in those who are re-experiencing the trauma and have the greatest overall distress. People with PTSD also often have other sleep disorders at the same time: half or more may have sleep-disordered breathing, and 33 to 76 percent may have excessive periodic leg movements.

The relation of disturbed sleep to PTSD is not well understood. Some studies have shown that people who have the most sleep disturbance in the short term after a motor vehicle accident are more likely to develop PTSD over the next six to twelve months. It may also be that having poor sleep before the trauma predisposes to its later development. In a study of survivors of Hurricane Andrew in the United States, those who reported disturbed sleep or nightmares in the month before the hurricane were significantly more likely to later develop PTSD. Similarly, in a study of police officers, a history of poor sleep helped predict lower scores on overall health and higher scores on bodily complaints independent of PTSD symptoms.

Although there is some reason to think, then, that disturbed sleep may predispose to PTSD there are also some data that total sleep deprivation immediately after a traumatic event may actually lower its emotional impact. In a study from the University of Oxford and the Karolinska Institute in Stockholm, volunteers were shown a film with upsetting content. One group was allowed to sleep afterward while the other was kept awake. Twelve hours later, the sleep-deprived group scored lower on a psychological measure of impact of stressful events, and had fewer intrusive emotional memories over the next six days. Since one of the possible functions of sleep is to help with the formation of long-term memories, one could speculate that preventing sleep immediately after a traumatic event somehow decreases the likelihood that the experienced trauma will become so deeply and emotionally lodged in memory.

Treatment of sleep difficulty in PTSD has at least two major goals. The first is to treat co-existing sleep disorders. In those who have sleep-disordered breathing for instance, appropriate treatment is associated with a decrease in PTSD symptoms in general. The medicine prazosin, which blocks one particular type of receptors for norepinephrine is often given by doctors to improve sleep and overall clinical status, and decrease nightmares though it is not specifically approved for this purpose ("off label use"). Limitations of prazosin can be excessive lowering of blood pressure. Many people with PTSD also have accompanying depressive or anxiety symptoms, which can be treated with appropriate medications. The SSRI antidepressants sertraline and paroxetine are often used by doctors for treating PTSD. Psychotherapy, particularly types known as imagery rehearsal and retraining therapy and cognitive behavioral therapy, can reduce symptoms. Even in treated people, however, though their overall symptoms may greatly improve, up to half may continue to have sleep disturbance in the long term.

SLEEP IN MEDICAL DISORDERS

Illnesses which cause disturbed sleep

A wide variety of medical illnesses can cause disturbed sleep, not only due to discomfort from pain but from directly altering sleep processes. Here we will describe some examples.

Hyperthyroidism

The thyroid gland can become overactive for a variety of reasons, such as a side effect of some medications, or of illnesses including Graves' disease and toxic goiter. Among the many symptoms can be tremor, oversensitivity to heat, rapid heart beat, and a kind of staring expression due to effects on the eyes. People with hyperthyroidism often describe insomnia, nightmares, and night sweats. The sleep EEG typically shows excessive amounts of N3 sleep, and indeed when looking at a sleep study, doctors are trained to consider the possibility of hyperthyroidism when this is noted along with a relatively higher heart rate. After hyperthyroidism is treated, it may take some time for the N3 sleep to return to normal amounts.

Night-time gastroesophageal reflux (GER)

In some people, acid from the stomach can move upward into the esophagus when they are lying down. This can lead not only to night-time heartburn, but also to chronic cough or chest pain not associated with heart disease. The resultant poor sleep can in turn make a person even more sensitive to esophageal pain. GER is also associated with an increased risk of sleep apnea and nocturnal asthma. Sometimes GER is silent, that is, without producing symptoms of heartburn, and it may represent a hidden cause of insomnia and daytime fatigue. GER can be assessed during a sleep study by inserting an electrode sensitive to acidity into the esophagus. Treatments for GER include lifestyle changes (quitting smoking, losing weight, avoiding spicy foods) and medications including antacids, or medicines that block acid production.

Chronic kidney disease

Chronic kidney disease (CKD) affects as many as one-tenth of the general population. Many of these people describe significant insomnia or sleepiness during the daytime. It turns out that up to half have sleep-disordered breathing; restless legs syndrome and periodic leg movements are also frequently present. Recognizing the sleep-disordered breathing clinically can be harder because these people are often thinner, are less likely to have a snoring history, and ultimately have a higher likelihood of central sleep apnea. A high portion of patients who require chronic peritoneal dialysis (in whom dialysis fluid is introduced into the abdominal cavity) develop sleep apnea, at least partially due to the effects of the bulk of this fluid on diaphragm function. It seems likely that a number of factors contribute to the high rate of sleep disorders in CKD. Some difficulties associated

MEDICINES THAT CAN DISTURB SLEEP

Not only medical conditions, but also the drugs used to treat them, can disturb sleep. Among these are prescription medications such as:

- Some asthma medicines that dilate the bronchi
- Beta-blocker medicines for high blood pressure and angina
- Nicotine patches
- Stimulants used for ADHD
- Steroids
- Thyroid medications
- Some SSRI antidepressants
- Some statin medications for high cholesterol
- Some medicines for Alzheimer's disease ("cholinesterase inhibitors")

Over-the-counter medication such as:

- Nasal decongestants
- Pain relievers or other products containing caffeine
- Some herbal preparations such as St. John's Wort

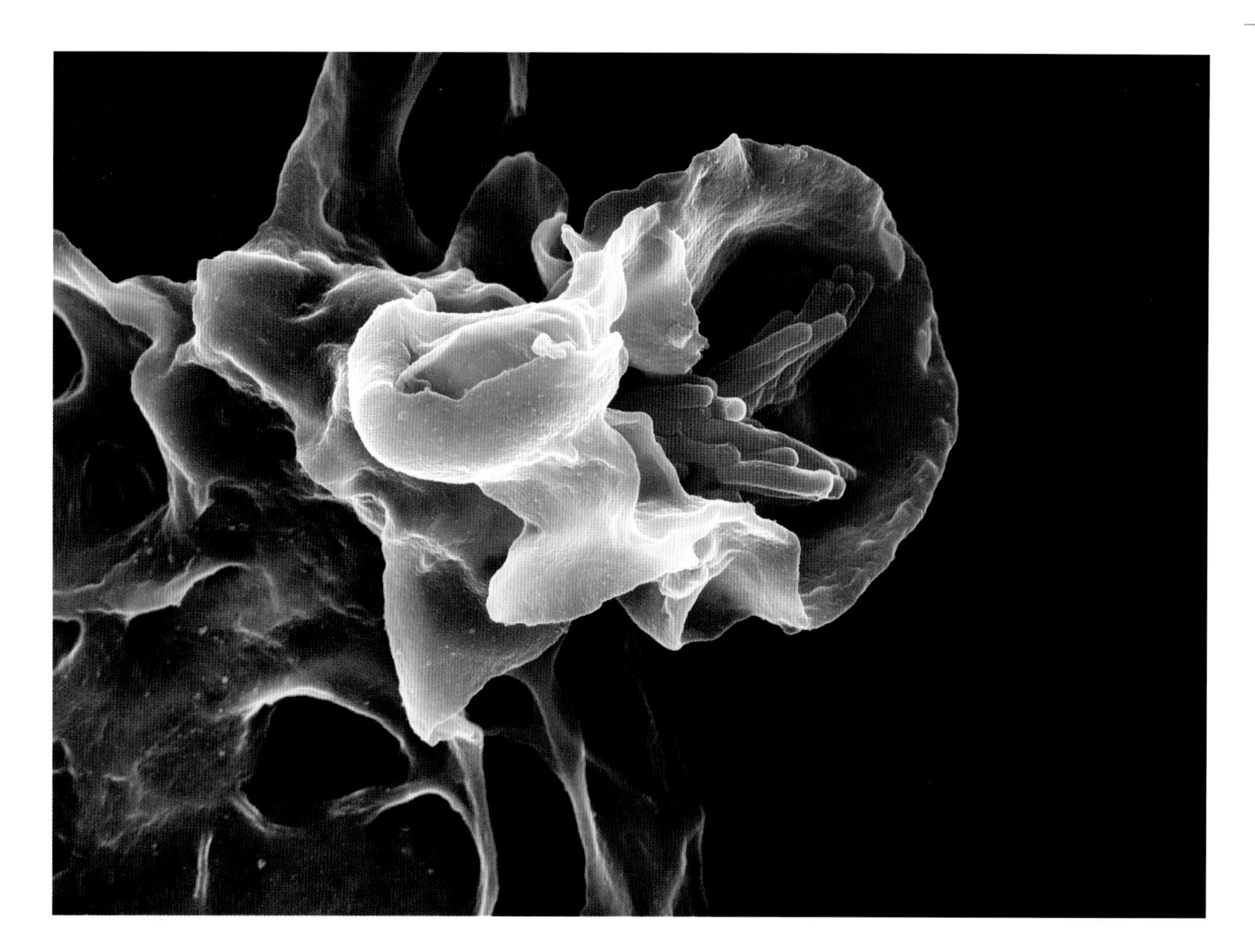

IMMUNE SYSTEM
A scanning electron micrograph of a macrophage consuming a bacterium. Macrophages are a type of white blood cell which (along with some other kinds of cells) also release cytokines, small proteins that modulate the immune response and affect sleep. The overlapping interaction of sleep and the immune system is an ongoing subject of research.

with CKD such as skin itching and mood changes may disturb sleep. The timing of hemodialysis procedures may play a part, particularly when performed on the early shift. Some of the chemicals not fully removed by dialysis, known as uremic toxins, may make a contribution. Many doctors treat the restless legs syndrome with some of the drugs typically used in RLS patients in general. Obstructive sleep apnea is managed by making sure dialysis is sufficient, and sometimes continuous positive airway pressure is used.

Infections and the role of cytokines

Sleepiness is a common response to acquiring a variety of infections, and indeed NREM sleep is often increased, while REM diminishes. The mechanism by which this occurs may be an increase in activity of a series of molecules involved in the immune system, known as cytokines. Both NREM sleep regulation as well as immune processes share some common compounds including interleukin-1 and tumor necrosis factor, which are produced by a type of white blood cell known as macrophages. The role of sleep in recovery from infections is a topic that continues to be explored. One study found that animals who manifest a large increase in NREM sleep after the onset of an infection are more likely to recover than those who do not.

SLEEP IN ALCOHOL DEPENDENCE

What effects does alcohol have on sleep?

The effects of alcohol on sleep depend on whether one has a history of alcohol dependence (excessive drinking for such prolonged times that withdrawal syndromes such as delirium tremens can occur on cessation). In people with no history of heavy drinking, alcohol can shorten the time taken to fall asleep. This benefit for sleep is questionable, however, as alcohol is broken down in the body relatively quickly, and a kind of "mini-withdrawal" may occur in the second half of the night, with sleep disturbed by multiple awakenings. REM sleep is decreased, and at higher doses slow-wave sleep may decline as well. Thus, although alcohol may help you go off to sleep, you usually pay the penalty of overall greater sleep disturbance. It's a common observation that people differ in their sensitivity to alcohol's effects. Animal studies suggest that a genetic component is involved, both in terms of the ability of the body to chemically break down alcohol, as well as in the brain's sensitivity to its sedating effects. The amount of prior sleep a person has had also influences how sedating alcohol can be. Not surprisingly, people who had less sleep prior to drinking experience a greater sedating effect from alcohol. The opposite is true for people who have had a larger amount of sleep. In one experiment in which volunteers stayed in bed for 10 hours per night for a week, when later given alcohol they did not fall asleep more quickly than when receiving a non-alcoholic drink.

The way that alcohol produces sleep is complex. In animal studies, injections of tiny quantities of alcohol into the medial (or median) preoptic area of the hypothalamus (see pages 36–7) induce sleep, suggesting that the hypothalamus may be one important location in this process. It seems likely that alcohol acts at the GABA receptor complex (see pages 158–9), as in animal studies drugs that block or increase activity of the receptor respectively inhibit and enhance sleep produced by alcohol.

Many people who have histories of alcohol abuse but are currently "dry" complain of very disturbed sleep. The laboratory confirms this, showing that sleep is very fragmented, and that N3 sleep is reduced. Indeed, one of the reasons people who in the past have abused alcohol may return to drinking is the belief that it will improve their sleep. In the very short term, there may be some truth to this; sleep latency becomes shorter, there is less sleep fragmentation, and amounts of N3 sleep rise. Of course, in the long term, this is the worst thing one can do in this situation, as resuming drinking perpetuates the sleep disturbance.

In alcohol-dependent people, withdrawal from drinking often results in a rebound increase in REM sleep. In extreme cases, this can be associated with alcoholic hallucinosis, in which there are active hallucinations in the daytime roughly 12 hours after the last drink, or delirium tremens 72–96 hours after the last drink, with confusion, agitation, hallucinations and overactivity of the sympathetic nervous system.

After the initial period of alcohol withdrawal has passed, many such individuals continue to have disturbed sleep up to two years after stopping drinking, reflecting the long-term changes chronic alcohol causes in the brain. It is difficult to know what medicines, if any, can be of help. Many doctors are reluctant to prescribe typical hypnotics (sleeping pills) because of their dependence-producing

potential. Some use the sedating antidepressant trazodone in an "off-label" manner, though it is not specifically recommended for this purpose. The experimental compound L-tryptophan, which is a chemical precursor of serotonin, has been reported to help normalize sleep, speculatively reflecting a correction of alcohol-induced alteration of biogenic amines (see pages 36–7) in the nervous system. Cognitive behavioral therapy for insomnia can be helpful.

ALCOHOL AFFECTS SLEEP QUALITY
The Offering to Bacchus (1720) by Michel Ange Houasse (1680–1730). This painting from Greek mythology shows the followers of Bacchus, the god of the vine, celebrating in his honor. Some are inebriated, some sleep. Alcohol has profound effects on sleep, which differ depending on whether one is only an occasional drinker or is alcohol-dependent. Although alcohol may aid in going off to sleep, any benefits are lost in the second half of the night when there is a mini-withdrawal process with many awakenings. Alcohol dependence can lead to long-term sleep disturbance which may not fully dissipate for prolonged periods after cessation of drinking.

Chapter Eight

INSOMNIA

Insomnia can be a long-term, difficult problem, involving not only discomfort at night, but adverse effects on daytime functioning. It can affect one's work, driving skills, and quality of life, and is associated with health conditions including high blood pressure, cardiovascular disease, and a propensity toward diabetes. It may also be associated with the later development of depression. The good news is that there are a number of treatments available. These include learning ways to change thinking and behaviors related to sleep, as well as medications. Each has its own set of benefits as well as limitations. They are also not incompatible with each other, and are sometimes combined. It is hoped that the information you will find here will make you more informed about treatments that are available as you discuss any sleep difficulties with your doctor.

WHAT IS INSOMNIA?

How can chronic insomnia affect your life?

Insomnia, like pain, is usually defined as a personal experience. In the case of insomnia, it is the feeling a person has of difficulty going to or staying asleep, awakening from sleep unrefreshed. This is accompanied by complaints of daytime difficulties such as fatigue or poorer functioning, even when there has been adequate opportunity to sleep at night.

The most common single symptoms of insomnia are having multiple awakenings during the night or unrefreshing sleep, but most people with insomnia have a combination, for instance having a hard time going to sleep as well as awakenings during the night. The type of insomnia symptoms a person has may change over time; there is some evidence that those with a single symptom such as difficulty going to sleep have a more variable pattern, while those with multiple symptoms (for instance trouble going to sleep plus awakenings during the night) may have more consistent patterns. In both the United States and the United Kingdom, roughly one third of the population complain of these night-time difficulties almost every night, and about one third of this group—10 percent of the population—feel that their insomnia results in problems in their daytime lives. These include fatigue, impulsiveness or irritability, trouble with memory, and difficulty in personal and professional relationships. People with these daytime symptoms have a higher rate of utilization of medical care facilities, including emergency room visits, doctor appointments, and phone calls. Sleep disturbance at least a few times a week is reported 1.4 times more frequently in women than men and, interestingly, is more common in parents than in those without children at home (66 percent vs 54 percent). Most studies have shown that the frequency of sleep complaints rises across the lifetime, and is highest in the elderly, although one reported the highest rate in adults aged 35 to 64, with a slight decrease in people over 65. There is a weak relationship with socioeconomic status, with a slightly lower rate of insomnia complaints the higher the socioeconomic level.

Insomnia, daytime functioning, and health

Insomnia is associated with many kinds of practical difficulties in one's daytime life. People with insomnia have lower scores on measures of quality of life. A study of US Navy recruits demonstrated that those who said they did not sleep well when entering the service were less likely to later be promoted or asked to reenlist, and generally did more poorly. Another indicated that prospective bank employees who mentioned sleep difficulties as part of an intake health screening were less likely to continue working there over time. A recent study of simulated driving for one hour on a highway showed that people with insomnia were more likely to stray out of their lane and had more difficulty with the car swaying instead of staying in a straight line. When asked about their perception of their driving skill at the time, those with insomnia did not differ from good sleepers, raising the worrisome possibility that they were

DAYTIME CONSEQUENCES OF INSOMNIA

Insomnia is not just a problem of sleeping at night. It is associated with a variety of daytime difficulties, including:

- Difficulties with concentration, mood, and energy
- Alterations in neuropsychological testing
- Decreased pleasure from family relationships
- Automobile accidents due to fatigue
- Absenteeism from work
- In the military, decreased promotions or invitations to re-enlist
- Increased healthcare consumption and costs

unaware of their impairment. Interestingly, those with insomnia had about the same amount of total sleep the night before the simulated driving test. This makes the point that insomnia is somewhat different from just plain sleep loss, a topic we will talk about later. People with insomnia describe less satisfaction in their personal relationships. They are also at a much higher risk to develop depression. This may be particularly true of people with insomnia who sleep less than 6 hours, who may be more vulnerable to the many complications of insomnia.

Insomnia is also associated with a number of health conditions including hypertension (high blood pressure) and altered metabolic function including insulin resistance (a risk factor for diabetes) or type 2 diabetes. There is a higher frequency of cardiovascular morbidity, particularly in people with insomnia who have significantly shorter sleep and higher levels of physiologic arousal such as elevated heart rates. The nature of this relationship is not well understood; it is possible that insomnia leads to these conditions due, for instance, to alterations in inflammation or arousal levels, or alternatively that they are associated but not causally linked.

UP ALL NIGHT
Insomnia is a very common condition, and can have significant impact on one's daytime life. It may come from a variety of causes. A number of treatments are available, including medications, changes in lifestyle, and non-medication talk therapies.

HYPERAROUSAL IN INSOMNIA

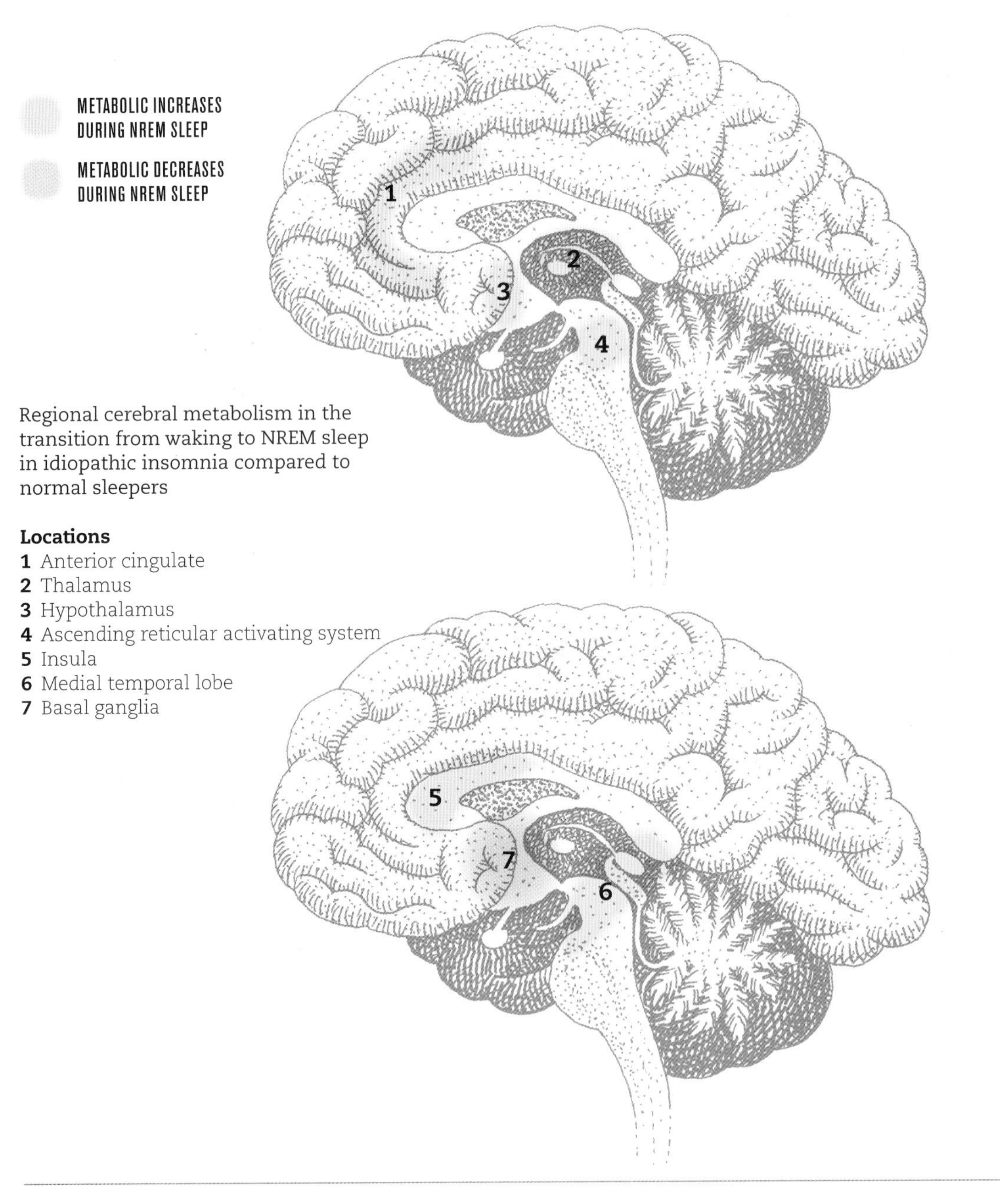

Regional cerebral metabolism in the transition from waking to NREM sleep in idiopathic insomnia compared to normal sleepers

Locations

1 Anterior cingulate
2 Thalamus
3 Hypothalamus
4 Ascending reticular activating system
5 Insula
6 Medial temporal lobe
7 Basal ganglia

METABOLIC RATE INCREASE

In this study of regional brain metabolism (measured by glucose consumption, as described in the photographs on pages 24–5), insomniacs show a smaller decline in metabolism compared to normal sleepers during the transition from waking to NREM sleep (areas 1–6). Of particular interest is the relatively higher rate in the thalamus, which might suggest prolonged degree of sensitivity to sensory information. One study using another technique known as SPECT (single-photon emission computed tomography) found decreased blood flow in the area known as the amygdala. The overall finding in the majority of areas is consistent with the notion that during the transition to sleep, there is hyperarousal in many areas of the brain in people with insomnia compared to good sleepers. Adapted from Desseilles et al (2008).

Hyperarousal

There are a number of ways of looking at the physiologic underpinnings of insomnia, but one particularly important one is the notion that insomnia is characterized by a state of hyperarousal (bodily or nervous system arousal which do not allow for sleep). This may share some features with "flight-or-fight" stress response. Early hints that this might be the case included the observation that the heart rate of poor sleepers in bed is higher than good sleepers, but a number of other physiological changes and brain imaging findings have been noted (see box). It is also possible that important changes in arousal in insomnia may be in specific regions of the brain. One recent imaging study suggests that in insomnia during NREM sleep there is a reduction in the normal disengagement from brain regions involving cognition, self-reference, and emotion, or conversely a decrease in engagement during waking. Many sleep scientists believe that although hyperarousal may be an important physiologic aspect of insomnia, a full understanding must take into account cognitive and behavioral processes as well.

Changing views of insomnia

Our views about insomnia are continuing to evolve. For many years, insomnia was considered to be primarily a symptom of some other illness, and the focus of sleep doctors was to find the underlying cause. The reasoning was, find and treat the underlying disorder, and the sleep disturbance will take care of itself. So, for instance, a clinician would look to see if someone with sleep difficulties had depression and, if found, would give antidepressants. The reasoning was that if the depression were successfully treated, the sleep disturbance would then take care of itself. Many other causes were looked for, including sleep troubles as side effects of another medication, or physiologically measured sleep disorders such as jerking of the legs in periodic limb movement disorder. Indeed when these are treated—for instance by stopping an offending medication that disturbs sleep—then sleep often improves. In more recent years, though, sleep doctors have come to realize that in addition to being the result of some other underlying condition, insomnia may also be a free-standing disorder in its own right, which needs its own treatment. Indeed, it has even turned out that treating sleep disturbance in the context of another illness may decrease its symptoms. Depressed patients, for instance, who are treated with sleeping medications in addition to an antidepressant may do better than those who receive only an antidepressant, and the pain in some kinds of arthritis may improve when the sleep is simultaneously treated. How can insomnia be both a symptom of some other illness and also be a problem in its own right? Actually, there are a number of areas in medicine in which this is the case. High blood pressure, for instance, may be due to a variety of underlying illnesses including kidney or endocrine disorders (which the doctor should look for as part of the assessment), but high blood pressure can also be an independent problem which needs treating, in the absence of other illnesses.

HYPERAROUSAL IN INSOMNIA

Some evidence of hyperarousal in people with insomnia:

- Increased basal metabolic rate
- Increased urinary levels of catecholamines (see page 34), a sign of sympathetic nervous system overactivity
- Increased body temperature
- Decreased respiratory-induced variation in heart rate in sleep (a sign of autonomic nervous system arousal).
- Increased beta activity on the EEG (a sign of arousal of the cerebral cortex)
- Increased brain metabolism in sleep, and less of a decline in brain arousal areas with sleep onset
- Increased arousal as measured on psychological tests

INSOMNIA DISORDER

Two major forms of insomnia

Insomnia may exist by itself, or in combination with other medical or psychiatric conditions. When it is freestanding, a number of factors can contribute to its presentation.

Adjusting to new situations

Adjustment sleep disorder refers to disturbed sleep in response to a specific stressor, such as an upsetting event or change in environment or illness. It is acute (short-term), lasting from a few days to a few weeks, and generally clears up when the stressor goes away.

Developing conditioned responses

This is a chronic (long-term) condition that can grow out of acute insomnia in some individuals. Initially a stressful event causes poor sleep. In some people, though, the disturbed sleep comes to take on a life of its own, persisting after the original stressor resolves. Sometimes the individual focuses on the sleep disturbance, rather than the stressor, and becomes highly concerned about the inability to sleep well. A kind of conditioning process occurs: the individual comes to associate the sleeping environment (the bedroom) and the act of going to bed with anxiety about not sleeping, and develops a kind of physiological arousal that cuts in at bedtime and which prevents sleeping. This may be accompanied by self-defeating behaviors such as going to bed too early or staying in bed for too many hours. It can be complicated by mistaken beliefs about sleep ("Most people get a full night's sleep," or "I'll never make it through the day if I don't get eight hours sleep").

One well-known way of looking at the development of chronic insomnia is the "3 Ps" model:

- Predisposition for poor sleep;
- Precipitating event (the acute stressor); and
- Perpetuating factors (behaviors such as going to bed too early, or mistaken beliefs about sleep).

When sleep studies are performed on these individuals, there is fairly good correspondence between what the person reports (such as a prolonged time to fall asleep) and what is seen on the sleep study (polysomnogram).

Discrepancies between personal experience and the polysomnogram

Paradoxical insomnia (sleep state misperception) is a less common condition (perhaps 5 percent of people with chronic insomnia) in which there is a poor correspondence between the significant degree of sleep disturbance that the person describes, and what is measured objectively on the traditional polysomnogram. A person might say, for instance, that (s)he is getting only 2–3 hours sleep every night, but the sleep study finds that physiologic measures of sleep indicate 6 hours. These people are absolutely convinced that they are getting little sleep and that their daytime lives suffer for it. What appears to be happening is that they have a kind of misperception of when they are physiologically asleep. One study that demonstrated this process was conducted by having people with and without insomnia go to bed in the laboratory. When the EEG indicated that they were asleep (a common marker would be that they are in N2 (stage 2) sleep ten minutes after the first appearance of sleep spindles), the investigator would go into the bedroom, awaken the subject, and ask: "What were you doing when I came into the room just now?" The normal sleepers would most likely answer, "Well, of course, I was sleeping." When the investigator asked the same question of the people with insomnia, who had appeared to be asleep and whose EEG showed N2 (stage 2) sleep, the response was often: "Well, I was lying here awake, or course." What seems to be happening, then, is that for reasons that are not well understood, there is a discrepancy between how these people perceive themselves (as awake) and their behavior and objective physiological measures (asleep). It is possible that as technology improves with advances in EEG and imaging, that this apparent discrepancy will be resolved.

THE COST OF INSOMNIA

In addition to the distress and discomfort it causes, insomnia also has an economic impact on society. An Australian study found that the costs of treating sleep disorders plus related expenses such as lost productivity, absenteeism, and sleep-related automobile accidents was over $7 billion annually. The total costs of insomnia alone have been estimated at $2 billion annually in France and variously from $30 billion to $107 billion in the United States.

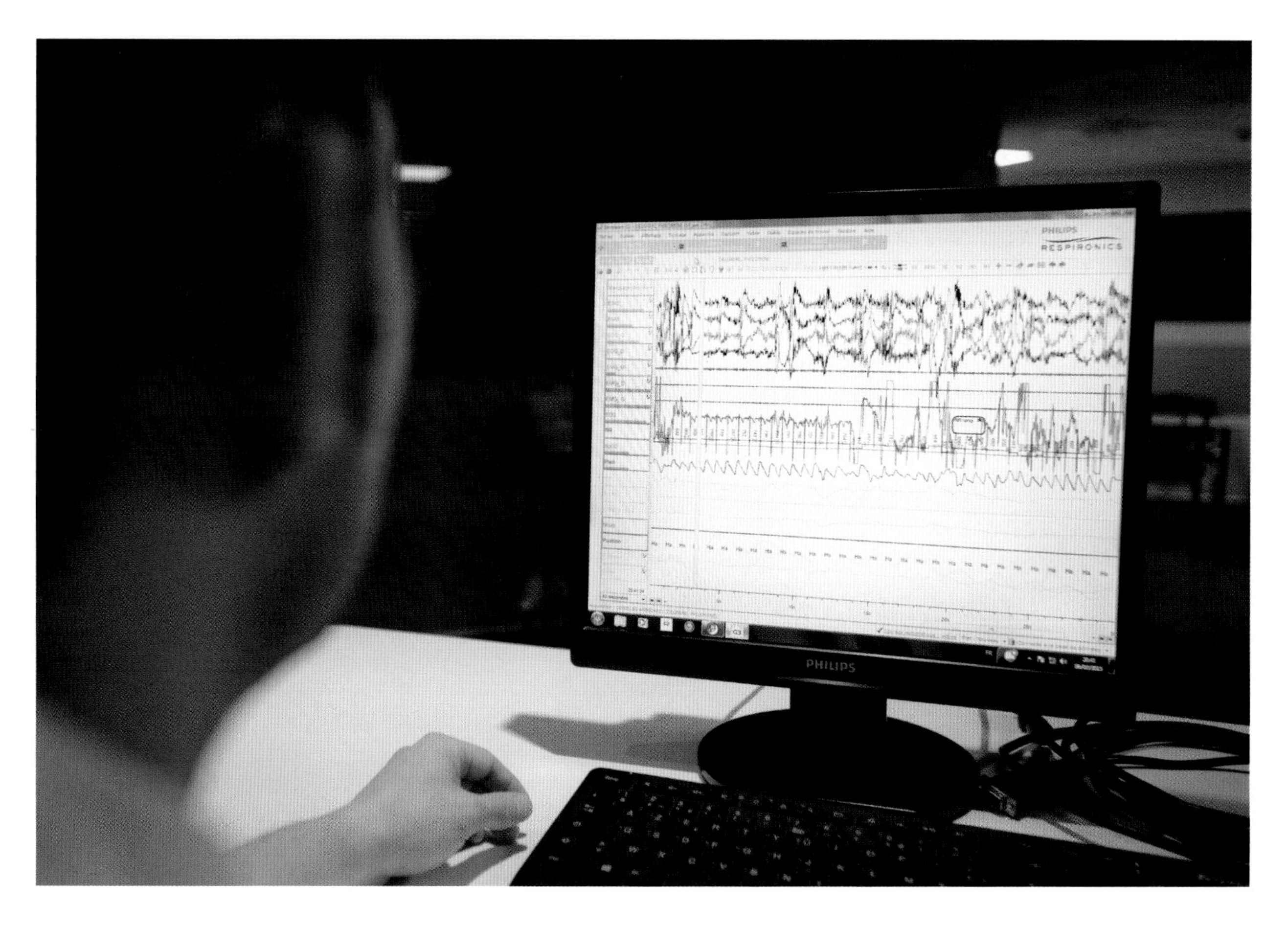

STUDYING INSOMNIA
In a French laboratory, an all night polysomnogram is performed to help determine the cause of disturbed sleep. In addition to looking for disorders of sleep physiology, it can be important to compare a person's subjective report of how (s)he slept with the objective measures of sleep found on the sleep recording.

Inadequate sleep hygiene

Inadequate sleep hygiene refers to a situation in which a person engages in behaviors that are not compatible with good sleep. Examples might include someone who keeps extremely irregular hours of going to bed and arising, takes excessive naps, or drinks coffee or engages in stimulating activities around bedtime. Again, there may be some element of susceptibility in certain individuals: the same behavior that disturbs sleep in one person may not do so in another. Later, when we get to the treatment of insomnia, we will look at a general set of common sense practices that may aid sleep, known as sleep hygiene. Also, it should be noted that if there is extreme excess of use of sleep-disturbing substances such as coffee or alcohol, the resulting sleep disturbance would fall into the category of "co-morbid" insomnia.

CHILDHOOD INSOMNIA

We often think of insomnia as an adult problem, but many children have sleep disturbances as well. One large study of schoolchildren in Belgium, ages 6–13, found that 62 percent had some form of sleep disturbance (defined as a caregiver describing difficulty at least three nights per week for the last six months). The most common were disorders of excessive sleepiness and difficulties in initiating and maintaining sleep.

Contribution of as-yet poorly understood factors

Some people have a poorly understood condition that goes back to childhood, which has sometimes been referred to as "idiopathic insomnia." It is not associated with a stressful precipitating event, and it continues on through adulthood, usually without any significant periods of sleeping well. There is no obvious history of conditioning factors such as are seen in psychophysiological insomnia, and sleep disturbance continues through periods of high or low stress or anxiety. It is not well understood, but may be associated with high levels of physiological arousal. Other forms of insomnia associated with childhood include limit-setting sleep disorder, in which the child stalls or will not agree to go to bed, requiring firm limit setting on the part of the parent, and sleep onset association disorder, in which the child becomes attached to inappropriate objects or behaviors. (S)he may, for instance, refuse to go to bed except under special conditions such as having the lights on or being in one particular place, or holding a particular object.

Insomnia in combination with other conditions ("co-morbid insomnia")

Often insomnia co-exists with another medical or psychiatric disorder, or with substance abuse. A person with arthritis, for instance, might be kept up by pain, or a depressed individual might spend sleepless nights dwelling on unhappy thoughts. Various forms of substance abuse may result in poor sleep as well. This can result from stimulating effects of some abused substances, from withdrawal symptoms while the drug is stopped, or from long-term effects which continue after stopping. A good example of the latter is alcohol dependence, in which sleep disturbance may persist for up to two years even after a person has stopped drinking, due to chemical changes induced in the brain by alcohol. Sleep can also be disturbed by a variety of prescription medicines (see box on page 140). Co-morbid insomnias are probably the most common situation, with only about 15 percent of people with sleep disturbance having an isolated insomnia complaint. The most common situation (40 percent) is insomnia in the presence of a psychiatric problem, typically anxiety disorders or depression.

Accompanying medical disorders occur in about 35 percent; these most commonly are hip conditions, chronic obstructive pulmonary disease, or heart conditions. About 10 percent of people with insomnia have an additional sleep disorder such as periodic leg movement disorder or central sleep apnea. Insomnia in people with circadian rhythm disorders is addressed in Chapter Four (see pages 82–5). For this reason, if a person with insomnia has received standard treatments but the symptoms persist for some time, it can be useful to have a sleep study and look for such conditions.

The way sleep researchers have thought about treating co-morbid insomnias has changed over the years. Formerly the general notion was "treat the underlying disorder, and the sleep will take care of itself." By this reasoning, if a person had insomnia accompanying depression, then give antidepressants and psychotherapy; as the mood starts to improve, so will the sleep. In more recent years, a new approach has been evolving. One study indicated that giving depressed patients a sleeping pill (eszopiclone) along with an antidepressant (fluoxetine) resulted in better antidepressant response. Another found that adding a sleeping pill to the regimen reduced the pain in rheumatoid arthritis patients. It appears, then, that treating the sleep difficulty can make it easier to manage the coexisting illness. Doctors find that it can be helpful, then, to use a sleeping pill in conjunction with the main medication to treat a medical or psychiatric disorder—in some cases, the benefit can go beyond just helping the sleep. Similarly, it appears that the non-medication approach to insomnia known as cognitive behavioral therapy can improve measures of depression as well as sleep.

TECHNIQUES TO IMPROVE SLEEP WITHOUT A PILL

Non-medication approaches to treatment

These techniques to aid sleep generally have the advantage of not having the types of side effects which must be considered when taking medication, and many sleep specialists consider them the place to begin in dealing with chronic insomnia. The most basic step usually taken is to review a person's sleep habits, and educate him/her in the practices of good sleep hygiene. Again, these are general principles. Some behaviors that disturb sleep in one person can cause no difficulty in another. For some people, for instance, keeping irregular bedtimes goes to the heart of their sleep disturbance, while others tolerate it with no difficulty. Napping during the daytime may be related to difficulty going to sleep at night for some, while for others it is a refreshing habit that does not interfere with sleep.

There are many individual therapies, as well as multimodal programs which combine them and such as cognitive behavioral therapy for insomnia (CBT-I). Some of the individual approaches include relaxation therapies, stimulus control, and sleep restriction.

Relaxation therapies

Some relaxation procedures focus on making the muscles less tense. This is done by an exercise in which one alternately tightens and then relaxes each major muscle group in the body in turn. When the musculature is relaxed, a person often experiences a peaceful kind of feeling; the idea is to learn to recognize this feeling, and ultimately try to induce it even without going through the muscle tensing and relaxing procedure. In the past, muscle biofeedback (a procedure in which a person learns to relax while being given real-time information from a machine that measures muscle tone) has been used for this purpose, but its effectiveness was never very well established and it is not used very frequently now.

For many people, excessive arousal is more in the mind than in the muscles. In this case, there are attention-focusing techniques including visualizing relaxing imagery or meditation. Here the goal is to focus the mind, and help prevent intrusive, worrisome thoughts that often accompany bedtime in insomnia. It may be that the old folk remedy of visualizing and counting sheep was a homegrown method of doing this. Focusing on these playful little creatures jumping over a fence might have helped prevent dwelling on more disturbing thoughts that make it difficult to fall asleep. Relaxation procedures can often be helpful. The only real negative is that if a person is very perfectionistic, then the effort of trying to relax can paradoxically be anxiety-provoking.

SLEEP HYGIENE MEASURES

- Keep a regular sleep schedule, going to bed and getting up around the same time each day.
- Have a bedroom environment conducive to sleep: quiet, dark, comfortable temperature (slightly cool is probably better than warm).
- Exercise during the day may aid sleep; exercise in the evening may be disturbing to sleep.
- Avoid heavy meals around bedtime. Do not use stimulating substances such as caffeine or nicotine near bedtime.
- Do not take alcohol to help sleep. Any benefits in going off to sleep are usually of less importance than the awakenings it produces in the second half of the night.
- Allow time in the evening to relax. Do not work right up until bedtime and then jump into bed expecting to sleep.

DAILY EXERCISE

There are a number of behavioral steps that one can take to help aid sleep (sleep hygiene). Among these is getting sufficient exercise. This needs to be done during the daytime, however; exercise in the evening can sometimes be disturbing to sleep. An important thing to remember about sleep hygiene is that these are general principles, and individuals differ in which combination of behavioral steps are helpful to them.

Stimulus control

The notion here is that a person with chronic insomnia forms associations between the physical setting of sleep—the bedroom—and the uncomfortable and anxiety-producing sense of being unable to sleep. Often this stimulus–response association is an unconscious one (remember Pavlov's dog salivating at the sound of a bell because of prior association of the bell and food?). The treatment goal is to strip the bedroom of all stimulus–response associations except for sleeping. A person is told, then, to remove all other behaviors from the bedroom: no watching television or surfing the internet, no bill paying, no thinking about one's problems or what one is going to do tomorrow. Another behavior which is not allowed is lying in bed awake: the instruction is, if you find yourself lying in bed unable to sleep for perhaps 20 minutes, then get up and go in another room and don't return to the bedroom until you feel sleepy.

MOBILE DEVICES AND SLEEP

The stimulus control treatment approach emphasizes the importance of stripping the bedroom of all behaviors aside from sleeping. One common activity in children and adolescents is the use of mobile media devices. A report in the United Kingdom indicates that 72 percent of children have at least one mobile device in the bedroom. Studies have shown an association of screen time with sleep disturbance, and of night-time use of media-related technology (often while consuming caffeine beverages) with daytime sleepiness in adolescents. Although these types of studies do not clearly show cause and effect, the association is of some concern. Another aspect of this issue is that melatonin secretion (see pages 80–1) is particularly sensitive to blue light, which is given off in significant amounts by mobile devices. There are now some apps to alter the color spectrum from mobile media and manufacturers are looking at ways to limit blue-colored emissions during the night-time hours.

Sleep restriction

The basis of sleep restriction therapy is that, regardless of the reason a person is not sleeping well, one of the complicating factors which helps perpetuate the problem is staying in bed for a much longer time than one actually sleeps each night. A person is asked how many hours (s)he typically stays in bed each night, and how many of those hours are actually spent sleeping. If (s)he reports being in bed 8 hours but only sleeping 5, then (s)he is requested to begin staying in bed only 5 hours. The goal is to make sleep more efficient by limiting the time available for it. When (s)he reports being asleep for at least 80 percent of these 5 hours, then the time in bed is allowed to increase in 20-minute increments until a longer and more efficient sleep is produced.

Cognitive therapy

The goal of cognitive therapy is to address mistaken notions and excessive preoccupations with sleep, both of which can heighten anxiety which contributes to poor sleep. Thus if a person is convinced that poor sleep is the reason behind all of his/her daytime difficulties in getting things done, it might be useful to consider whether there are other issues involved, for instance conflicts in the family or at work. Another thought that can be addressed with cognitive therapy is the notion that sleep is totally out of the person's control because of a "chemical imbalance" or "inherited sleep problem." Such a faulty attribution needs to be identified and eliminated through logical persuasion so that the individual will make behavioral changes to address the insomnia. While indeed sleep is very important, excessively focusing on it may make things harder.

Other beliefs that need to be reconsidered might be the thought that if a person gets less than 7 hours sleep on a particular night, (s)he will be completely unable to function the next day. It's useful to try and develop realistic expectations about sleep, and not to expect a miraculous change for the better. While the hope and expectation is that things will get somewhat better, there will always be some nights of poor sleep, and one must learn to accommodate them.

Multi-modal therapy, or CBT-I, combines these various techniques, and is thought by some sleep doctors to be the most effective in terms of long-term benefits. Often it is done in weekly meetings over a period of six to eight weeks. There is some evidence that even one or two visits to a therapist in the context of a primary care clinic may be of some help. There are also CBT programs available online. A study of one of them known as SHUTi (www.myshuti.com) found that after a six-week course of treatment, about half the people who took it were sleeping better a year later, compared to about one quarter of a control group. It is not yet clear how to predict in advance which people are more likely to benefit from this approach.

CBT-I THERAPY

CBT-I involves implementing a variety of approaches to dealing with sleep difficulty. A typical set of techniques might include:

- Considering thoughts and behaviors that lead to excessive arousal
- Stimulus control therapy
- Sleep restriction therapy
- Addressing any body clock (circadian) issues
- Considering the effects of foods and substances on sleep

HOW DO SLEEPING PILLS WORK?

Pharmacological approaches to treatment

Prescriptions for hypnotics (sleeping pills) number about 10 million annually in the United Kingdom and 60 million annually in the United States. The story of how sleeping pills were first developed and how they evolved begins with Adolph von Baeyer (1835–1917). He was a fascinating scientist, a student of Robert Bunsen (1811–99)—remember the Bunsen burner with which so many of us heated foul-smelling chemicals in high school chemistry class?—and the organic chemist August Kekule (1829–96), who, legend has it, dreamed of a snake biting its tail, and from this deduced the idea of ring-like chemical structures. He later went on to win the Nobel Prize in Chemistry for inventing indigo blue and other dyes, but at this point in his life, Baeyer was a poor graduate student in Ghent, assigned the task of combining the chemicals urea and malonic acid into a ring structure. At last, after many failed attempts, on December 4, 1864, he was successful, and went out on the town to celebrate. When he neared the tavern, he found it already crowded. The patrons are said to have been soldiers from a nearby military base, who had their own reason for celebration, the day of St. Barbara, the patron saint of the artillery. It was apparently quite an evening, and sometime during the revelry, the words "Barbara" and "uric acid" are said to have come together in the inebriated student's mind to form the name of his new discovery, barbituric acid. It opened the door to a wide range of chemical compounds, and was the basis from which a host of sedatives, sleeping pills, and anesthetics evolved.

Barbiturates

After the development of barbital (United States) (barbitone elsewhere; trade name Veronal), the first clinically used barbiturate, these compounds became the most widely prescribed sleeping pills from the early 1900s until the 1960s. Though potent at inducing sleep, they had many drawbacks, including potential for abuse and dependence, toxicity in overdose (as few as 10 pills could be lethal), stimulation of breakdown of other drugs by the liver, and depression of respiration. For these reasons, they are (rightfully) very rarely used any more for sleep. We have mentioned them here because experiences with them have influenced the way people have thought about sleeping pills in subsequent years, and they also make a kind of benchmark against which to measure the improvements in subsequent generations of medicines.

Benzodiazepines ("Valium-like" drugs)

Beginning in the 1960s, a new class of tranquilizers known as benzodiazepines became available, and by 1970 ones recommended for sleep came along. They act by binding to a specific site (sometimes informally called the "Valium receptor") which is associated with the receptor for the inhibitory neurotransmitter GABA (gamma-aminobutyric acid; see pages 36–7). Because of many advantages over the barbiturates, they rapidly became the dominant tranquilizers and sleeping pills. Among those available currently in the United Kingdom are diazepam, temazepam, loprazolam, and nitrazepam, and in the United States temazepam, lorazepam, diazepam, and others. Their many advantages over older drugs included decreased toxicity in overdose (though they could indeed be lethal especially when combined with other compounds such as alcohol), a more moderate respiratory depressant effect, and a lower, but still present, likelihood of abuse compared to barbiturates. Unlike barbiturates, they do not

HOW BENZODIAZEPENES WORK

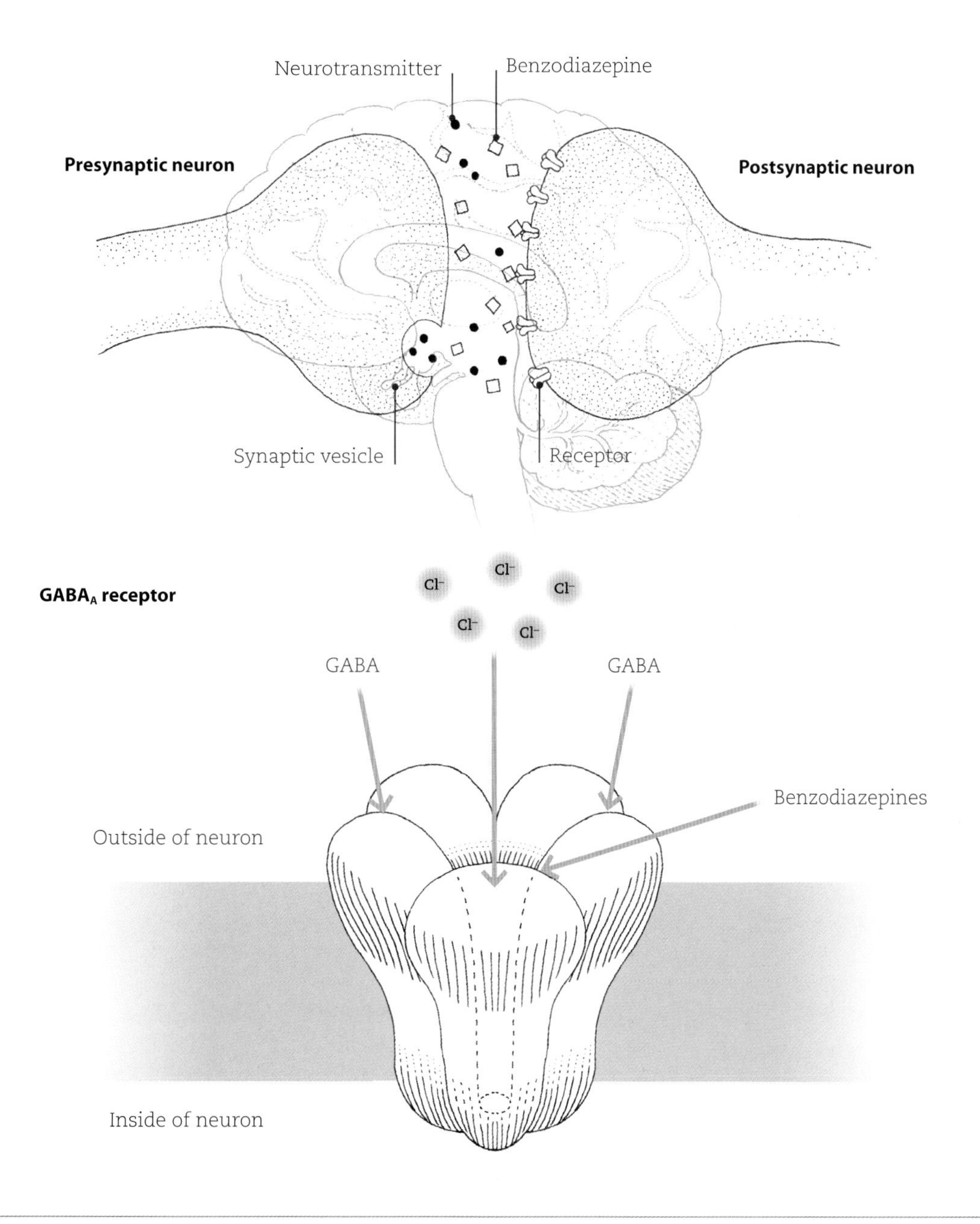

THE GABA$_A$-BENZODIAZEPINE RECEPTOR COMPLEX

This is a complex structure which is found on the surface of neurons sensitive to GABA, comprised of five proteins (which come in multiple forms) built around a central channel which passes through the cell membrane. Included in the overall structure are places at which GABA and benzodiazepines ("Valium-like" drugs) can bind, known as recognition sites. Activation of the receptor leads to the flow of negatively charged chloride ions (Cl^-) from outside the cell to inside the cell through a channel (indicated by the dotted lines on the above graphic). The resulting increased negative charge inside the cell leads to a state in which the neuron is less active. The receptor complex is also affected by other agents, including ethanol, barbiturates, neurosteroids, and the anesthetic propofol. There are also two more types of GABA receptor, known as GABA$_B$ and GABA$_C$, which have other functions.

stimulate the liver to increase the breakdown of other drugs. Benzodiazepines prescribed for sleep such as flurazepam were found to reduce time to fall asleep after two to three nights of administration, and to increase total sleep time. Unlike barbiturates, they had only mild effects on REM sleep, though they suppressed slow-wave sleep.

Beyond the benzodiazepines

In 1988, zolpidem tartrate, the first of a new class of sleeping pills, was introduced into Europe, and subsequently came to the United States in 1993. Though their chemical structure is very different from benzodiazepines, they act by binding to the receptor site affected by benzodiazepines, with more precision in affecting the portion related to sleep. These drugs for sleep, which currently include zolpidem and zopiclone in the United Kingdom, and zolpidem, zaleplon, and eszopiclone in the United States, rapidly became popular because of several perceived advantages. Among these is specificity of action; they are thought to be more likely to affect sleep and to have relatively fewer effects on other processes affected by benzodiazepines such as muscle tone and memory. Unlike the benzodiazepines, they do not significantly suppress slow-wave sleep, though the clinical meaningfulness of this is not clear. Zolpidem itself reduces sleep latency (helps a person to go to sleep sooner), but has relatively little effect on total sleep time and awakenings in the second half of the night, which led to the development of an extended release preparation. The longer-acting zopiclone has more potent effects on total sleep time. In more recent years, new forms of administration of zolpidem have become available, including one which dissolves under the tongue, and an oral spray.

These non-benzodiazepine sleeping pills (sometimes called "GABA agonists," as they enhance the action of GABA) appear on paper to have many advantages over the benzodiazepines, but in real-world use these are perhaps not so clear. A survey in England of over 700 patients found that they reported no clear advantages in effectiveness or adverse effects. A review of twenty-four studies with a combined group of over 3,000 patients found few consistent differences in effectiveness or side effects between the newer medicines and the benzodiazepines. There is one area in which advantages to the newer drugs appeared in these types of studies—a review of doctors' records in Germany found there were only about one-third the number of mentions of misuse. It is important to remember, however, that they can be misused and abused, and are classified as drugs with this potential.

TERMS USED IN THIS SECTION

Sleep latency: The time from when the lights are turned out until the beginning of sleep. A sleeping pill that helps you to go to sleep sooner may be said to reduce sleep latency.
Sleep maintenance: Refers to the continuity of sleep after initially falling asleep. A medication that reduces the number of awakenings during the night might be said to improve sleep maintenance.
$GABA_A$-benzodiazepine receptor complex: A complex protein structure on the surface of some neurons which includes a place to which benzodiazepines attach (the benzodiazepine recognition site, or "Valium receptor"), an area to which GABA and some other drugs attach, and a channel for chloride ions. The benzodiazepines and "GABA agonist" sleeping pills attach to this structure and enhance the activity of the inhibitory neurotransmitter GABA (see pages 36–7).
Receptor agonist: A substance which binds (attaches) itself to a receptor, and promotes its activity. For instance, the sleeping pill ramelteon (see page 162) is a receptor agonist for some melatonin receptors. In contrast to an agonist, an antagonist binds and blocks the action of a receptor.

USE OF HYPNOTICS

To put the use of hypnotics in perspective, a 2016 study found that about one in six Americans reported filling a prescription for a psychiatric drug in 2013, mostly for long-term use (filled at least three times, or having taken it for at least two years). The most common type of medicine was antidepressants, followed by tranquilizers and sleeping pills. In this latter group, the sleeping pill zolpidem was one of the most common, followed by the antidepressant trazodone which is often used for sleep (see "Off label use of medicines and OTC sleep aids" on page 163).

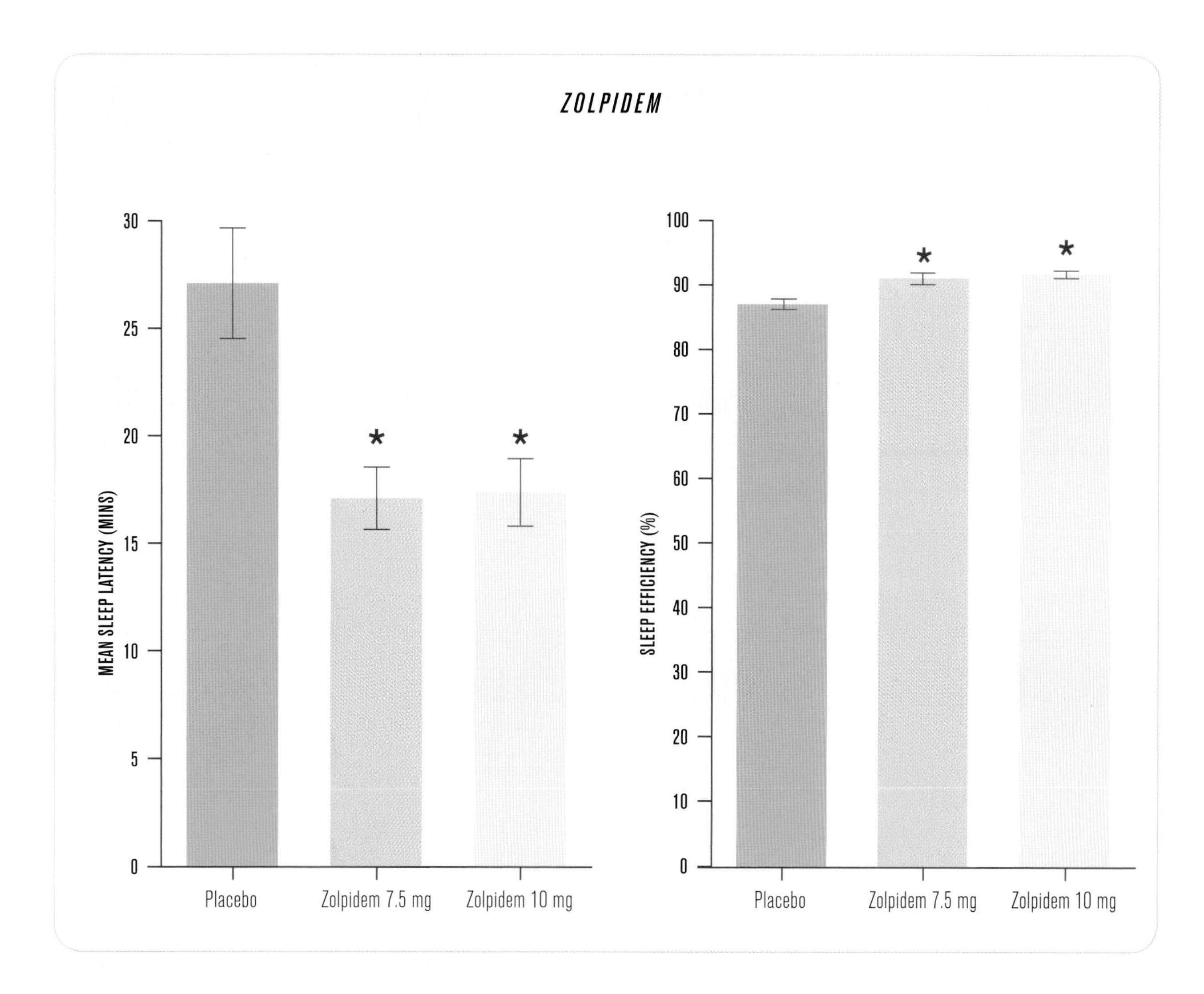

EFFECTS OF ZOLPIDEM

A study by Thomas Roth et al. (1995) of zolpidem in normal volunteers in whom the "first night effect" (sleep disturbance from sleeping for the first time in a laboratory) was used as a model of insomnia. Zolpidem in doses of 7.5 mg and 10 mg reduced sleep latency (the time to fall asleep) and improved sleep efficiency (the percentage of the time in bed occupied by sleep). The asterisk indicates that there is less than one in a thousand likelihood that this difference from placebo was due to random chance.

Other sleeping pills

As we mentioned earlier, all the sleeping medicines which we covered so far produce their effects by acting on components of the gamma-aminobutyric acid (GABA) receptor complex. In recent years, newer drugs have been developed which operate in other ways.

Ramelteon

Ramelteon, available in the United States but not Europe, acts by binding to two types of the receptor for melatonin (see pages 78–9), about 17 times more tightly than melatonin itself. It appears to have no dependence-producing potential, and is one of the two sleeping pills (along with doxepin) which are not classified as having restricted use because of possible dependence. It is a very short-acting agent, with little or no effect on performance the next day. It does not have respiratory depressant qualities. It appears to help people go to sleep, but only after taking it for about a week, with little or no benefit on total sleep time. Perhaps for these reasons it is not very widely prescribed.

Doxepin

Doxepin represents a case of a new use for an old drug. Available for decades as an antidepressant in the United States and Europe, a very low dosage preparation became available for sleep in the United States in 2010. It has a variety of chemical actions including effects of norepinephrine (noradrenaline) and serotonin (see pages 36–7), and very potent antihistaminic properties. (Indeed there is a preparation of doxepin as a skin salve for itching.) Low-dose doxepin's sleep benefits are primarily on sleep maintenance (reducing waking time during the night), which produces increased total sleep time.

SLEEPING PILL USE

In this study of Americans over 20 years old, in the period 2005–2010, one in eight with trouble sleeping said they were using prescription sleep aids. About 4 percent of all adults had used a prescription sleep aid in the last month. In general, use increased with age and amount of education. This may be part of a broader trend in increased use of psychotropic medicines in the elderly. One recent study found that the number of older Americans taking at least three psychotropic drugs doubled from 2004 to 2013. This is of special concern due to the risk of drug interactions and because the elderly are potentially more susceptible to side effects.

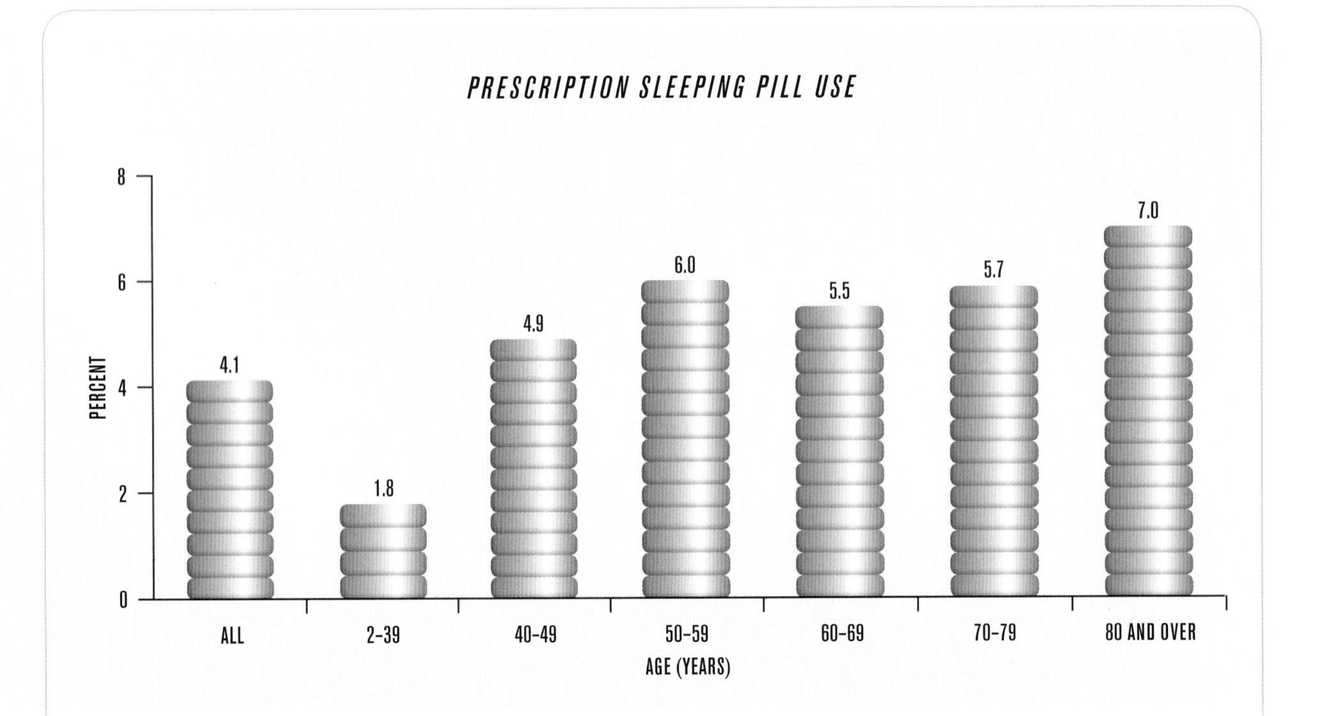

Time to fall asleep is not consistently reduced. The tricyclic antidepressants, the drug class which includes doxepin, have a number of concerns in the higher dosages used for depression. These include toxicity in overdose, confusion, effects on the heart, dry mouth, and difficulty with urination; these are potentially present but possibly of somewhat less concern at the very low doses employed for sleep. Doxepin appears to be relatively free of dependence-producing properties, and generally does not carry related restrictions.

Suvorexant

Available in the United States and Japan but not Europe, suvorexant acts by blocking the receptors for the wake-promoting orexin/hypocretin peptides in the hypothalamus (see pages 36–7). A relatively short-acting compound, the lowest dose has relatively little effect on daytime functioning, and minimal effects on respiration. Its effectiveness involves both reduction of sleep latency and aiding sleep maintenance, although more so the latter. Because of inhibiting orexin activity (and since a decrease in orexin function has been found in the illness narcolepsy) there has been concern some patients may develop sleep paralysis, hypnogogic hallucinations, or cataplexy-like symptoms, but these have generally not been observed in controlled studies. Suvorexant carries restrictions for dependence-producing properties similar to the benzodiazepines and non-benzodiazepne GABA agonists, although available data suggest minimal risk.

Off label use of medicines and OTC (over-the-counter) sleep aids

Another pharmacologic approach that is often used is to prescribe medications which are intended for other purposes, but which are sedating and are thought to help sleep. This practice is known as "off-label" prescribing. Perhaps the most common medication used in this manner is the antidepressant trazodone. One thing to be aware of in using off-label medicines in this manner is that they have often not been tested for the purpose at hand. This is true of trazodone, for which (despite its very wide use) there is limited systematic data in people with insomnia without accompanying depressive illness. One of the few head-to-head comparisons done compared its nightly use with zolpidem. Fundamentally, the study found that at the end of one week, both drugs shortened sleep latency, though zolpidem did so more powerfully. By the end of week two, only the zolpidem continued to show a benefit to sleep latency. Although the systematic data are limited, many physicians have the clinical impression that trazodone can be helpful for insomnia. It also has one positive feature compared to the benzodiazepines and newer GABA agonists in that it does not carry restrictions related to dependence. On the other hand, it is important to remember that as an antidepressant, trazodone has its own set of side effects, including sedation, effects on the heart, falls and hip fractures, and priapism (painful and often prolonged medically serious erections).

In contrast to the situation in people with insomnia without depression, people with major depression accompanied by insomnia may find their sleep helped by trazodone (as well as actual sleeping pills). In one study of this group it did not benefit sleep latency, but reduced the number of awakenings and improved total sleep time.

Our understanding of the appropriate time to use sedating antidepressants such as trazodone in insomnia without depression is evolving. Current guidelines by the American Academy of Sleep Medicine suggest that, in patients who are not responding well to treatment, doctors try at least two conventional sleeping pills such as GABA agonists or ramelteon before using sedating antidepressants. When weighing the possible benefits and risks of such off-label prescribing, it is good to note a Canadian study of off-label use and side effects across all classes of medications. It was found that when drugs are used off-label, there is a 44 percent greater chance of an adverse drug events occurring compared to on-label use.

CONCERNS ABOUT SLEEPING PILLS

Balancing the benefits and risks

Though sleeping pills can often provide relief for sleepless nights, like all medications there is potential for a variety of unwanted effects, most of which are more common at higher doses. It is important for you to discuss these with your doctor if (s)he chooses to prescribe something to help you sleep. A partial list includes:

Abuse and misuse

As mentioned earlier, most sleeping pills (the exceptions include ramelteon and doxepin) are considered to have potential for abuse, particularly among people with histories of substance abuse. When recommended doses are stopped abruptly, insomnia may return, but any serious discontinuation effects aside from transiently disturbed sleep are rare. Abrupt discontinuation of very high doses (considered misuse) can cause much more severe symptoms including significant sleep disturbance for a few nights, agitation, confusion, and even seizures. For this reason, many doctors try to minimize discontinuation effects by tapering down from higher doses over a number of nights before discontinuing the drug completely.

Tolerance (losing effectiveness over time)

Virtually all sleeping pills are recommended for short-term use, and have primarily been tested in that manner. Most studies of benzodiazepines were for one month or less, and the longest formal testing for the newer non-benzodiazepines include a study of eszopiclone for six months under rigorous study procedures. Suvorexant has been examined for three months. In general, then, the effectiveness when taken over the longer term is not well documented in formal studies. Although the preponderance of evidence to date is that tolerance is not a major concern with currently available agents, the possibility of losing effectiveness might be kept in mind.

Mortality and suicide

There have been some epidemiological studies which show an association of sleeping pill use with a higher mortality rate. It is difficult to assess their significance. One factor, of course, is that in the majority of cases insomnia does not occur alone, but rather in combination with other medical conditions. Most of these studies were done in the years when benzodiazepines, or even older drugs, were the most widely used sleeping pills. In one prominent study, the higher mortality was associated not with prescription sleeping pills per se, but rather with drugs of other types taken to aid sleep, which often included pain killers.

A recent review of the literature confirms that hypnotics are associated to some degree with a higher risk of suicidal thoughts or suicide. The meaning of this is not entirely clear, however, as most studies did not adequately consider the possibility of co-existing depression or other psychiatric disturbances. Sleeping pill use is associated with parasomnias (see pages 130–3), which in rare cases can involve suicidal behavior. On the other hand, sleeping pill use in conjunction with antidepressants may aid in helping depression in some situations (see pages 136–9), and ongoing research is looking into whether they may potentially lower risk of suicide in this setting.

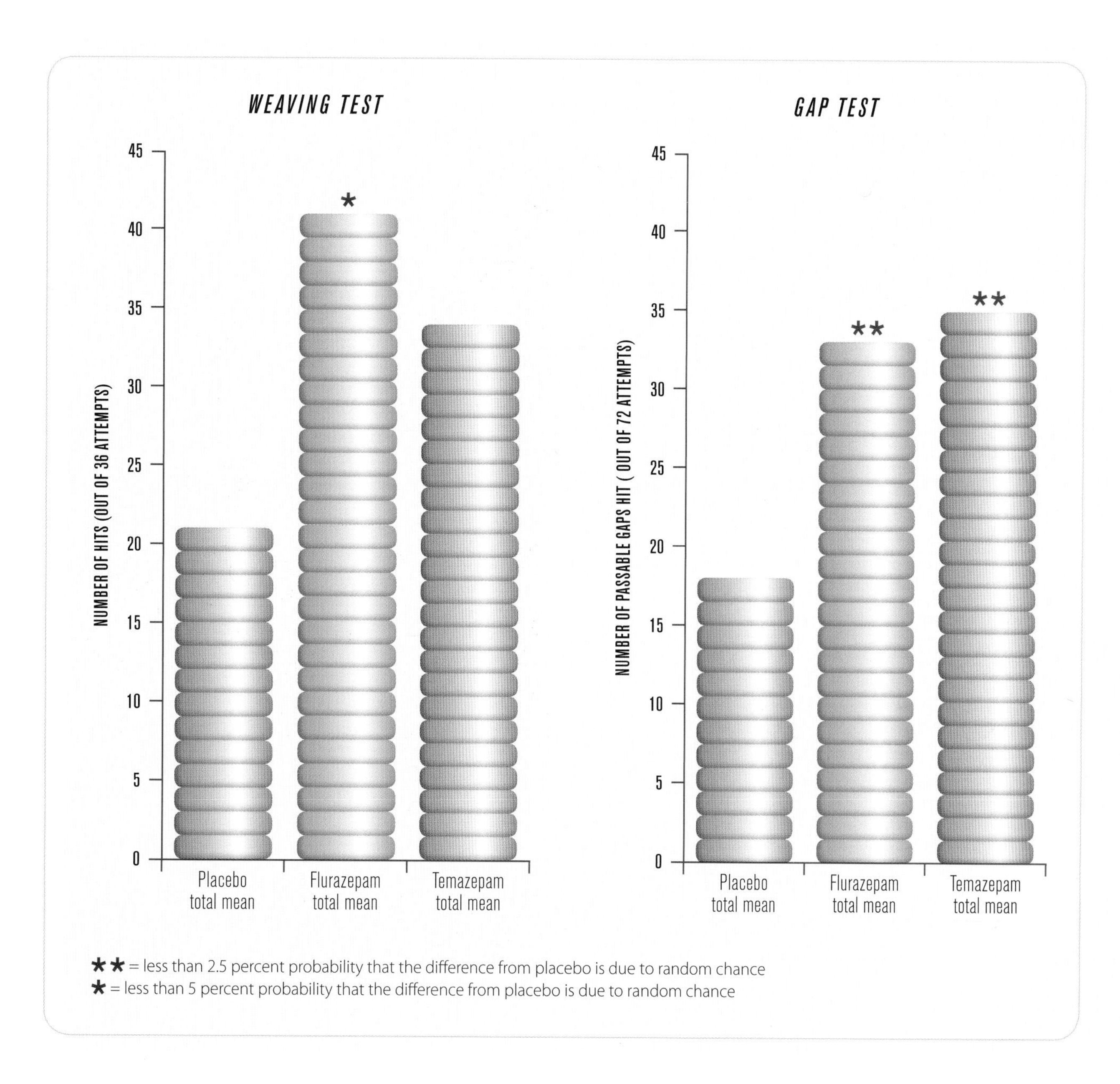

THE MORNING AFTER

One concern about hypnotics is that, although they are taken to benefit sleep, the effects of some can potentially influence one's daytime life. This is particularly true of longer-acting agents. In this 1982 study by Betts and Birtle, 12 healthy women volunteers who drove regularly were given placebo, the long-acting sleeping pill flurazepam 15 mg, or the somewhat shorter-acting temazepam 20 mg on different nights separated by one week, and had their driving tested the next morning 12 hours later. In the first task (left), they drove a Datsun saloon in a task in which they had to weave back and forth between bollards. On mornings following flurazepam, they had a greater number of incidents of hitting the bollards. In the second task (right), they were presented with narrow gaps between bollards and had to choose whether they were wide enough and then drive through them. In the morning following both drugs, they had a higher rate of striking the bollards in gaps that were wider than the car.

Interaction with other drugs and toxicity in overdose

Sleeping pills can also interact with other medications a person may be taking. One concern is greatly increased sleepiness, or confusion, when another sedative is taken at the same time. In overdoses, the toxicity of many of these drugs is greatly increased when taken in combination with alcohol.

Daytime sedation

Although sleeping pills are taken for the purpose of promoting night-time sleep, sometimes their effects continue on into the daytime, resulting in sleepiness or decrements in performance. This is particularly true for the longer-acting drugs, which in some cases can remain, and even accumulate, in the body during the daytime. The active compound into which flurazepam is converted in the body lasts so long that only half is removed after 47–100 hours. If taken nightly (every 24 hours), it can be seen how it ultimately would accumulate in the body. The biggest concern is that, although there may be daytime decrements in alertness, thinking, and skills (possibly including driving), a person is not necessarily aware of being impaired. Fortunately, the most commonly used sleeping pills are short-acting with less of a chance of significant daytime impairment.

Respiratory depression

The benzodiazepines represented a step forward when compared to the older barbiturates in that they produced substantially less inhibition of respiration, and the newer non-benzodiazepines have generally improved on this further. Measures of respiration are typically unchanged in healthy individuals. In people with underlying respiratory disorders such as sleep apnea or chronic obstructive pulmonary disease, however, this can be a problem, for instance potentially causing the sleep apnea to be more severe. Although there have been claims that some individual agents have few respiratory effects, it is best to be very cautious about the use of sleeping pills in the presence of respiratory illness. Of course this is true in other medical conditions as well. In people with significant liver or kidney disease, for instance, the ability to break down and remove drugs body can be decreased.

UNUSUAL NIGHT-TIME BEHAVIORS

There has been some association of the use of hypnotics with the rare occurrence of complex night-time behaviors such as getting up and driving or eating. In medication-induced sleep-related eating disorder, there is often no memory of the incident in the morning. Cases have been reported in which, over time, the increase in food consumption has led to substantial increases in weight and even the development of sleep apnea. The topic of nocturnal eating is complex, and medication-related behaviors need to be distinguished from eating disorders in which a person is fully awake during the episode and may have similar behaviors during the daytime as well.

Falls

Clinically, there is always a concern that a person taking sleeping pills can fall down when getting up during the night. Here the data are a little unclear. At least one prominent study indicated that it is insomnia itself which is associated with falls, and that the additional use of sleeping pills did not increase the risk further. In the absence of definitive information, the safest course is to view falls as a potential problem that needs to be considered.

Disinhibition reactions

Some drugs such as benzodiazepines are occasionally associated with outbursts of inappropriate anger or behavior. This occurs more when they are used as daytime anxiety drugs, and less so when they are used as sleeping pills. This is presumably due to the drug's suppression of processes by which higher brain centers in the cerebral cortex normally give "pause for thought" before speaking or acting on emotional impulses.

Amnesia

Most sleeping pills have been associated with anterograde amnesia, that is, inability to remember things that have happened after taking the medication. The degree to which this happens is difficult to assess, as the act of going to sleep itself has amnesic properties. (People who wake up for a few minutes in the middle of the night, for instance, often do not remember the awakening.) There have been rare reports of "global amnesia," often in travelers after taking sleeping pills during transcontinental flights. In these cases, there often appear to be multiple factors at work including jet lag, sleep deprivation, and concomitant use of alcohol.

Unusual behaviors during sleep

Sometimes people taking sleeping pills may engage in sleepwalking or more complex behaviors including going to the refrigerator for food, or driving, with no memory of these events the next day. This can happen in people who have never taken sleeping pills before, or those with a history of prior medicine experience. The likelihood of such events is increased by the concomitant use of alcohol or other sedatives. Although there have sometimes been claims that specific drugs are more likely to produce this side effect, the data are not clear and it seems prudent to consider it a possibility with any sleeping pill.

Allergic and other reactions

Allergic reactions are of course possible with any medication. A rare, but particularly alarming reaction is angioedema, in which the tongue or larynx becomes swollen, resulting in trouble breathing or in the throat closing. Symptoms suggestive of angioedema should be considered very serious and lead to immediate medical attention.

Balancing benefits and risks

In summary, sleeping pills, like all medications, can have a variety of adverse effects. Of course, it should also be considered that, as we described earlier, untreated insomnia carries its own set of detrimental effects on one's daytime life (though it's not clear that they are substantially reversed by sleeping pills). What this means, then, is that you and your doctor need to balance the potential benefits and risks associated with sleeping pills when coming to a decision about whether to try them or alternative non-medication treatments.

It should also be mentioned that sleeping pills are sometimes combined with non-medicine therapy. The studies of combination treatment have varied in results, but in general have not shown this to be more beneficial than either treatment alone. There have also been comparison studies of short-term nightly sleeping pill use with cognitive behavioral therapy. In general they suggest that sleeping pills act more quickly, but that cognitive behavioral therapy produces longer-term benefits after the cessation of treatment. So far there have not been any studies of cognitive behavioral therapy with long-term nightly use of sleeping pills. Similarly, there are few studies about the relative risks and benefits of taking sleeping pills on a non-nightly basis, even though this is how they are commonly used in practice. Most studies have been for three months or less, using the newer non-benzodiazpines, and have suggested no loss of effectiveness when taken in this manner, but our knowledge of longer-term use is lacking.

NO PRESCRIPTION NEEDED

Over-the-counter sleep aids and herbal preparations

In addition to prescription medications, there are of course a variety of sleep aids that can be purchased without prescription ("over-the-counter" or OTC). Generally these fall into two groups, those containing sedating compounds, typically antihistamines, and herbal products. The use of melatonin as a sleep aid is mentioned in Chapter Four (see page 78).

Antihistamines

Diphenhydramine and doxylamine are the most common antihistamines found in OTC sleep aids. As will be recalled from Chapter One, a histaminergic center in the hypothalamus promotes wakefulness, and it is thought that by blocking the histamine type 1 receptor, drugs such as diphenhydramine dampen this process. It is a first generation antihistamine with sedating properties; there is also a second generation including compounds such as loratadine which do not enter the brain as readily and are generally not used for sleep. Studies of the effectiveness of diphenhydramine have varied. In one study normal volunteers reported no effects on sleep, and measures of their motor activity during the night actually showed that they were more restless. Other studies in poor sleepers have sometimes shown improvement in reports of sleep latency and total sleep, and a study of older people found improvements only in the number of awakenings.

It is important to remember that even though diphenhydramine can be bought without prescription, it has a side effect profile just like any other medication. Some people have reported dizziness or ringing in the ears, diarrhea, or constipation. Particularly worrisome is use in the elderly, in whom it can potentially cause confusional states. It has also been associated with transient withdrawal sleep disturbance. During daytime use it may impair memory and learning processes, as well as making one sleepy. At least one study has suggested that the sleepiness decreases after a few days.

Herbal preparations

Probably the most common herbal preparations marketed for sleep are valerian extracts. They come from the root of *Valeriana officinalis*, a perennial herb growing in Europe, and cultivated in the United States and Japan. As a plant extract, it is actually a combination of many different chemical compounds. How they might aid sleep is not clear, but one compound slows the chemical breakdown of GABA, while another binds weakly to the benzodiazepine recognition site on the GABA receptor (see pages 36–7 and pages 158–9). Studies of efficacy have had mixed results. One study of elderly poor sleepers in the sleep laboratory found no change in sleep latency or waking time, an increase in N3 (slow-wave sleep) and decrease in N1 (stage 1) sleep, but the subjects themselves reported no difference in their sleep. Another study of people with mild insomnia found no changes in sleep stages, but a shortening of sleep latency beginning on day 28. The subjects themselves reported some benefit on their sleep beginning on day 14. One major review of valerian studies concluded that there is insufficient evidence to recommend valerian for improving self-reports of sleep.

Some people taking valerian can experience headaches, dizziness or restlessness, stomach disturbances, and even sometimes insomnia. It can cause daytime drowsiness which in principle might affect driving. Studies looking at safety have generally

VALERIAN PLANT EXTRACTS

Extracts from the valerian plant have been used for hundreds of years to aid sleep. These extracts are complex mixtures of many substances, which potentially could affect the brain in different ways. Scientific studies of the possible benefits of valerian on sleep have had mixed results and overall are largely inconclusive.

been limited to one month, so the long-term effects are unknown. It can also interact with certain medicines or other herbal products, for instance increasing sleepiness from other sedatives. As with many herbal products, the formal testing has been limited, so that the most appropriate dose has not been well determined.

It is also important to remember that herbal preparations in general are not manufactured with the same degree of regulatory scrutiny that is customary for prescription medicines. At the time of writing, the US Department of Justice had filed criminal and civil actions against 117 people and companies for issues including contamination of products or misleading claims.

In summary, there are pharmacologic alternatives to prescription sleeping pills. In the case of OTC sleep aids, there is evidence of very mild benefits with sedating antihistamines. However, these medications have their own side effect profiles which need to be weighed in balancing possible benefits and drawbacks. There is little systematic evidence that herbal preparations such as valerian significantly aid sleep, and any small benefits may take two weeks or more to appear. Herbal products, too, have side effects. Just because something can be bought without prescription, or is of natural origin, is not a guarantee of safety.

GLOSSARY

ADRENOCORTICOTROPIC HORMONE (ACTH) A polypeptide hormone released from the anterior pituitary gland, which stimulates the release of cortisol from the adrenal glands.

AGONIST In the context primarily used in this book, refers to a receptor agonist, a substance which binds to a receptor on neurons and facilitates the receptor's function.

AMINERGIC A broader term for the biogenic amines, including serotonin, norepinephrine (noradrenaline), and dopamine.

APNEA *See* **OBSTRUCTIVE SLEEP APNEA**.

AUDITORY AROUSAL THRESHOLD The volume of a sound required to wake a person up.

AUTONOMIC NERVOUS SYSTEM (ANS) Part of the peripheral nervous system that regulates a number of involuntary bodily functions, including for instance heart rate, blood pressure, digestive processes, and function of the pupil of the eye. It is in turn composed of two opposing divisions, the sympathetic and parasympathetic nervous systems.

BARBITURATES An older class of sedatives and sleeping pills (hypnotics), now rarely used for sleep because of their potential toxicity and the possibility of dependence.

BENZODIAZEPINES Valium-like sedative/hypnotics. Since their development in the 1960s, they largely replaced the more toxic barbiturates, and were the dominant sedative/hypnotics until the development of the newer GABA agonist hypnotics.

BRAIN STEM An evolutionarily older part of the brain situated at the base of the brain, just above the spinal cord. The brain stem is comprised of three divisions: the midbrain, pons, and medulla. Nerve centers in the brain stem are involved in a variety of processes including sleep/waking, but also for instance temperature, cardiac and respiratory activity, blood pressure, and swallowing. Most of the cranial nerves which regulate sensation and motor activity of the face come from the brain stem. Nerve signals to and from the upper portions of the brain to the spinal cord pass through the brain stem. Inside the brain stem are important groups of neurons (nuclei) containing acetylcholine, the biogenic amines, and other neurotransmitters which are involved in regulating sleep and many other functions.

CATAPLEXY One of the symptoms of the disorder narcolepsy–cataplexy, involving a sudden loss of muscle tone in the major weight-bearing muscles, and sometimes the head and neck, which occurs in association with emotion.

CENTRAL As in central hypersomnia or central apnea, refers to the central nervous system (the brain and spinal column). Thus central sleep apnea refers to a kind of sleep apnea in which the brain periodically fails to send messages to breathe to the diaphragm.

CENTRAL NERVOUS SYSTEM In vertebrates the central nervous system (CNS) comprises the brain and spinal cord. The other major portion of the nervous system is the peripheral nervous system, outside the brain and spinal cord, which connects them to the rest of the body.

CENTRAL SLEEP APNEA One of the two forms of sleep apnea (episodes of cessation of breathing during sleep). In central sleep apnea, the brain periodically fails to send signals to the body to breathe. On a polysomnogram, there is cessation of nasal airflow and, in contrast to obstructive sleep apnea, there are no respiratory efforts by the diaphragm and chest wall. Ultimately changes in blood chemistry cause a reflex resulting in resumption of breathing.

CHOLINERGIC NEURONS Neurons that use the neurotransmitter acetylcholine. Example: neurons of the laterodorsal tegmental nucleus (LDT) and pedunculopontine tegmental nucleus (PPT) which are the cholinergic "REM-on" centers in the reciprocal interaction model described on pages 38–9.

CONSCIOUSNESS Our experience of self and the world. It has been described as a paradoxical state in which a person is simultaneously a subject who can experience things and an object perceived by the self.

CONJUGATE EYE MOVEMENTS Eye movements that keep the angular relationship of both eyes, making it possible to fix on a single object and track a moving object.

DEPRESSION Usually in this context refers to the mental disorder known as major depressive disorder, a common disorder in which a person experiences symptoms including depressed mood, loss of interest or pleasure, feelings of guilt or low self-worth, disturbed sleep or appetite, low energy, and poor concentration. These symptoms last at least two weeks, and often much longer. Other related conditions include persistent depressive disorder (dysthymia) which lasts at least two years, seasonal affective disorder which appears primarily in the winter, psychotic depression, and the depression which appears as part of bipolar disorder.

DOPAMINE A neurotransmitter which plays an important role in alertness, pleasure, and rewarding behaviors, as well as functions in the peripheral nervous system. Dopamine-containing neurons are concentrated in midbrain nuclei of the brain stem. They have highest activity in waking and REM sleep, and dopamine is released in the cortex more in waking than sleep. The central nervous system illness Parkinson's disease is thought to be related to a loss of dopaminergic neurons in the midbrain.

ELECTROENCEPHALOGRAM (EEG) A recording of the brain's electrical activity. This activity is picked up by electrodes, small disks of metal attached to the scalp, and is then filtered and amplified electronically, and presented on paper or a computer screen. EEG data may be interpreted visually, or analyzed electronically. As used here, it is primarily an important part of a sleep recording or polysomnogram. In contrast are waking clinical EEGs, which (although there is substantial overlap) are often oriented more to detection of epilepsy and other disorders, and localizing particular EEG signals.

ELECTROMYOGRAM (EMG) Recordings of the electrical activity of muscles. In a sleep study, these are often done by recording from two electrodes placed on the skin over the anterior tibialis muscles on the lower legs.

ELECTROOCULOGRAM (EOG) Recordings of eye movements. In a sleep study, these typically record lateral eye movements, from electrodes placed near the outer canthi of the eyes (the corner of the eyes where the upper and lower eyelids come together).

GABA (GAMMA-AMINOBUTYRIC ACID) The most widespread inhibitory neurotransmitter in the brain. In general it lowers the level of excitability of nerve cells (neurons). In the context of this book, GABA acts by binding to the GABA recognition site of the $GABA_A$-benzodiazepine receptor complex.

GABA AGONISTS This is a common name for a class of sleeping pills introduced in the late 1980s and early 1990s, including zolpidem, zopiclone, and zaleplon. They are thought to act by binding as agonists at the benzodiazepine recognition site of the $GABA_A$-benzodiazepine receptor complex.

$GABA_A$-BENZODIAZEPINE RECEPTOR A complex protein structure on the surface of some nerve cells (neurons) that includes binding sites for the inhibitory neurotransmitter GABA, benzodiazepine (Valium-like) drugs, and an ion channel for chloride ions. Its function is also influenced directly or indirectly by a variety of sedative compounds including ethanol, barbiturates and the anesthetic propofol. When the receptor complex is stimulated, the chloride channel opens, allowing negatively charged chloride ions to pass from the outside to the inside of neurons, dampening their activity.

GABAERGIC NEURONS Use **GABA**. In terms of sleep, GABAergic neurons in the basal forebrain and anterior hypothalamus, for instance, fire more rapidly during NREM sleep than in waking. GABAergic neurons inhibit cholinergic neurons in the basal forebrain which are involved in arousal.

GLUTAMATE Amino acid glutamate is the excitatory neurotransmitter found throughout the brain, and in particular in the pontine and midbrain reticular formation. Application of glutamate to a number of brain sites leads toward wakefulness. It is involved in a variety of processes including memory and cognition.

GLUTAMINERGIC NEURONS Use glutamine. Example: Neurons descending from the brain stem to the spinal cord are involved in the mechanism of muscle relaxation during REM sleep.

GLYMPHATIC SYSTEM is a functional waste clearance system mediated by glial (non-neuronal) cells in the brain, which helps remove toxic substances from the nervous system.

HYPERSOMNIA Excessive sleepiness.

HYPOPNEA A partial apnea, present when nasal airflow is decreased by at least 30 percent, and accompanied by a drop in arterial oxygen saturation of at least 4 percent.

HZ (HERTZ)' Measurement of electrical waves in cycles per second.

IDIOPATHIC When an illness is called idiopathic, it means that the cause is not known. This is the case, for instance, for "idiopathic hypersomnia," a type of insomnia that is not well understood.

INSOMNIA The subjective experience of difficulty going to sleep, staying asleep, or awakening unrefreshed.

JET LAG is a condition resulting from rapidly crossing time zones in east–west travel, so that the body's internal rhythms are out of sync with the conditions of the new local environment.

MONOAMINES In the context of this book, a group of neurotransmitters including norepinephrine, serotonin, dopamine, and histamine.

NARCOLEPSY A sleep disorder involving very significant excessive sleepiness; in the condition known as narcolepsy–cataplexy, it is accompanied by the symptom cataplexy (see above), and often sleep paralysis, hypnogogic hallucinations, and disturbed nocturnal sleep.

NEURON An important type of nervous system cell which possess the qualities of irritability (excitability) and conductivity (ability to conduct impulses). Extensions known as dendrites receive impulses from other cells, while axons in turn pass impulses to other neurons.

NEUROTRANSMITTER A type of neurochemical produced inside neurons and released at synapses (the junction between an axon—see above), and an adjacent neuron. The neurotransmitter "bridges the gap" and conveys an impulse to the receiving neuron. A neurotransmitter's signal may be either excitatory or inhibitory.

NIGHTMARES AND NIGHT TERRORS A person who has nightmares awakens from sleep with a vivid memory of a very disturbing dream, which can involve fear but also other emotions such as anger or disgust. These events tend to occur in the second half of the night, and are associated with REM sleep. Night terrors involve awakenings, often with screaming and intense fear as well as autonomic arousal (rapid heart beat and respiration), but in contrast to nightmares with little memory of a specific dream, and little memory of the event the next morning. They tend to be associated with stage N3 (slow-wave) sleep.

NON-RAPID EYE MOVEMENT (NREM) SLEEP is, collectively, sleep stages N1, N2, and N3, previously known as stages 1–4. There are distinct electroencephalographic and other characteristics seen in each stage. NREM sleep, then, is comprised of all stages of sleep aside from REM sleep.

NORADRENALINE *See* **NOREPINEPHRINE**.

NORADRENERGIC NEURONS Use norepinephrine. Example: the locus coeruleus in the brain stem, which represents part of REM-off neurons in the reciprocal interaction model of the REM–NREM cycle, and other processes including the stress response.

NOREPINEPHRINE A biogenic amine neurotransmitter, playing an important role in a number of physiologic processes including sleep-waking regulation and the stress response. It also acts as a hormone when released into the bloodstream from the adrenal glands.

OBSTRUCTIVE SLEEP APNEA One of the two forms of sleep apnea, in which breathing periodically ceases during the night, due to a functional obstruction in the upper airway. On a polysomnogram there is cessation of nasal airflow, while the diaphragm and chest wall continue to make efforts to breathe. Ultimately, changes in blood chemistry produce a reflex resulting in resumption of ventilation. Obstructive sleep apnea produces a sense of disturbed sleep and daytime sleepiness, and is associated with a variety of health consequences.

PARASOMNIAS Disorders involving undesired or unpleasant experiences during sleep. Among these are disorders of arousal from NREM sleep such as night terrors and sleepwalking, and REM parasomnias such as nightmares.

PARASYMPATHETIC NERVOUS SYSTEM Part of the autonomic nervous system. It tends to quiet the body down, sometimes referred to as the "rest and digest" process. It tends to dominate activity during NREM sleep.

PERIPHERAL NERVOUS SYSTEM That part of the nervous system that lies outside the brain and the spinal cord.

POLYSOMNOGRAPHY (PSG) Sleep study recordings designed to measure a wide variety of physiologic functions, including EEG, electromyogram, arterial oxygen saturation, air movement from the nose and mouth, electrooculogram, electocardiogram, movement of the chest wall and diaphragm, and leg movements.

POSITIVE AIRWAY PRESSURE (PAP) is a major form of treatment for obstructive sleep apnea. It is basically a fan in a box, with a hose leading to a flexible mask placed over the nose. There are many variations, including devices which have differing pressures when breathing in or out, and ones which sense the amount of air pressure needed and adjust accordingly.

POSITRON EMISSION TOMOGRAPHY (PET) IMAGING A nuclear medicine, functional imaging technique that creates 3-D images of metabolic processes in the body. In one form, for instance, a radioactive tag is attached to glucose molecules, which are injected into the body. The PET apparatus detects positively charged particles coming from this chemical, and creates a map of the rate of glucose utilization in different areas.

POST-TRAUMATIC STRESS DISORDER (PTSD) A condition associated with having experienced stressful or frightening events. Along with a variety of other symptoms, about half of people with PTSD suffer from disturbed sleep, and 50–70 percent have nightmares.

RAPID EYE MOVEMENT (REM) SLEEP A stage of sleep characterized by low voltage, mixed frequency activity on the EEG, loss of muscle tone in the major weight-bearing muscles, rapid eye movements, irregularity of pulse and respiration, and the experience of dreaming. PET scan studies show activation of limbic and paralimbic areas (involved in emotions, drives, memory, and other processes) and, during periods of eye movements, activation of areas involved in arousal, attention, and emotion. On a typical night's sleep in a young adult, the first REM period may appear roughly 90 minutes after sleep onset, and there may be four to six episodes of REM throughout the night.

RECOVERY SLEEP Sleep which follows a prior period of sleep deprivation.

RESTLESS LEGS SYNDROME (RLS) A disorder in which there is an uncomfortable urge to move the legs, often accompanied by a "creepy-crawly" or tingly feeling, usually occurring at night or at rest, and relieved by movement.

RETICULAR ACTIVATING SYSTEM (RAS) A diffuse network of cells, running upward from the brain stem, which respond to sensory input, and bring a general message of stimulation to the cerebrum via the thalamus and a more anterior pathway through the hypothalamus basal forebrain. In classic studies in the 1940s, it was found that electrical stimulation of parts of the RAS led to awakening in sleeping animals.

SEROTONIN (5-HYDROXYTRYPTAMINE) A biogenic amine neurotransmitter, involved in regulation of many processes including sleep, mood, sexual function, temperature, bowel function, blood clotting, and bone density.

SEROTONERGIC NEURONS Neurons that use serotonin as a neurotransmitter. A major center of serotonergic neurons is the paired dorsal raphe nuclei in the brain stem. Ascending fibers from the dorsal raphe go to a wide variety of brain areas including the hypothalamus, basal forebrain, thalamus, striatum, and most of the cortex. The dorsal raphe nuclei are considered to be part of the "REM-off" neurons in the reciprocal interaction model of regulation of the REM–NREM cycle.

SHIFT WORK DISORDER (SWD) A significant minority (14–32 percent of night shift workers and 8–26 percent of rotating shift workers) can develop what is known as shift work disorder (SWD). Symptoms can include significant insomnia and sleepiness, even though the person involved may have a predictable sleep schedule, adequate available time in bed, and no other sleep disorders. Other symptoms can include difficulty concentrating, depression, and decreased energy. It is associated with medical disorders such as ulcers, and a higher rate of accidents.

SLEEP HYGIENE A set of guidelines for aiding good sleep, involving practices related to sleeping, dietary considerations, and the sleep environment, which are listed in Chapter Eight (see page 154). These are general principles, and all recommendations do not always fit all people.

SLEEP LATENCY The time from when the lights are turned out until the beginning of sleep. A sleeping pill that helps you to go to sleep sooner may be said to reduce sleep latency.

SLEEP MAINTENANCE Refers to the continuity of sleep after initially falling asleep. A medication that reduces the number of awakenings during the night might be said to improve sleep maintenance.

SLOW-WAVE SLEEP (SWS) Sleep stage N3, also stages 3 and 4 in the Rechtschaffen-Kales nomenclature. A deep stage of sleep characterized by the presence of high amplitude, slow (0.5–4 cycles per second) EEG waves.

SYMPATHETIC NERVOUS SYSTEM (SNS) Part of the autonomic nervous system, in the waking state it is associated with preparing the body for emergencies and rapid action, among other things dilating the pupils, increasing heart rate, reducing the rate of digestion, and inhibiting urination. In the context of this discussion of sleep and waking, it is associated more with REM sleep and arousal.

SELECTED BIBLIOGRAPHY

AKERSTEDT, T. AND NILSSON, P.M. "Sleep as restitution: an introduction." *Journal of Internal Medicine* (2003) 254: 6–12.

ANCOLI-ISRAEL, S. ET AL. "Sleep in the elderly: normal variations and common sleep disorders." *Harvard Review of Psychiatry* (2008) 16: 279–286.

ASCHOFF, J. ET AL. "Re-entrainment of circadian rhythms after phase shifts of the Zeitgeber." *Chronobiologia* (1975) 2: 23–78.

BOYAR, R. ET AL. "Synchronization of augmented luteinizing hormone secretion with sleep during puberty." *New England Journal of Medicine* (1972) 287: 582–586.

CALEM M. ET AL. "Increased prevalence of insomnia and changes in hypnotics use in England over 15 years." *Sleep* (2012) 35: 377–384.

CARSKADON, M.A. AND DEMENT, W.C. "Sleep loss in elderly volunteers." *Sleep* (1985) 8: 207–221.

COBLE, P.A. ET AL. "EEG sleep of normal healthy children. Part 1: Findings using standard measurement methods." *Sleep* (1984) 7: 289–303.

CORMAN, B. AND LEGER, D. "[Sleep disorders in elderly]." *La Revue du Praticien* (2004) 54: 1281–1285.

COSTA, G. "Shift work and occupational medicine: an overview." *Occupational Medicine* (2003) 53 (2003): 83–88.

EDINGER, J.D. ET AL. "Does cognitive-behavioral insomnia therapy alter dysfunctional beliefs about sleep?" *Sleep* (2001) 24: 591–599.

FINAN, P.H., QUARTANA, P.J. AND SMITH, M.T. "The effects of sleep continuity disruption on positive mood and sleep architecture in healthy adults." *Sleep* 38 (2015): 1735–1742.

GOLDBERG S. *Clinical Neuroanatomy Made Ridiculously Simple*. Medmaster, 2014.

GOLDMAN-MELLOR, S. ET AL. "Is insomnia associated with deficits in neuropsychological functioning? Evidence from a population-based study." *Sleep* (2015) 38: 623–631.

GREEN, D.J. AND GILLETTE, R. "Circadian rhythm of the firing rate recorded from single cells in the rat suprachiasmatic brain slice." *Brain Research* (1982) 245: 198–200.

GUILLEMINAULT, C., LEE J.H. AND CHAN, A. "Pediatric obstructive sleep apnea syndrome." *JAMA Pediatrics* (2005) 159: 775–785.

HILLMAN, D.R., SCOTT MURPHY, A., ANTIC, R. AND PEZULLO, L. "The economic cost of sleep disorders." *Sleep* (2006) 29: 299–305.

JIANG, F. ET AL. "Sleep and obesity in preschool children." *Journal of Pediatrics* (2009) 154: 814–818.

JOHAR, H., RASMILA K., THWING EMENY, R. AND LADWIG, K-H. "Impaired sleep predicts cognitive decline in old people: findings from the prospective KORA age study." *Sleep* (2016) 39: 217–226.

JUN, L., SHERMAN, D., DEVOR, M. AND SAPER, C.B. "A putative flip-flop switch for control of REM sleep." *Nature* (2006) 441: 589–594.

KRYGER, M.H., ROTH, T. AND WILLIAM C. DEMENT, W.C., eds. *Principles and Practice of Sleep Medicine*, Fifth Edition. St. Louis: Elsevier, 2011.

KUSHIDA, C.A. ET AL. "Effects of continuous positive airway pressure on neurocognitive function in obstructive sleep apnea patients: the apnea positive pressure long-term efficacy study." *Sleep* (2012) 35: 1593–1602.

LIPFORD, M.C. ET AL. "Associations between cardioembolic stroke and obstructive sleep apnea." *Sleep* (2015) 38: 1699–1705.

MCCARLEY, R. "Neurobiology of REM and NREM sleep." *Sleep Medicine* (2007) 8: 302–330.

MENDELSON, W. *Human Sleep: Research and Clinical Care*. New York: Plenum Press, 1987.

MOORE, T.J. AND MATTISON D. R. "Adult utilization of psychiatric drugs and differences by sex, age and race." Published Online: December 12, 2016. doi:*10.1001/jamainternmed.2016.7507*

MORRIS, C.J. ET AL. "Circadian system, sleep and endocrinology." *Molecular and Cellular Endocrinology* (2012) 349: 91–104.

PORCHERET, K., HOLMES, E.A., GOODWIN, G.M., FOSTER, R.G. AND WULFF, K. "Psychological effect of an analogue traumatic event reduced by sleep deprivation." *Sleep* (2015) 38: 1017–1025.

PRATHER, A.A., JANICKI-DEVERTS, D., HALL, M.H. AND COHEN, S. "Behaviorally assessed sleep and susceptibility to the common cold." *Sleep* (2015) 38: 1353–1359.

REIFMAN J., KUMAR K., WESENSTEN N.J., TOUNTAS N.A., BALKIN T.J., RAMAKRISHNAN S. "2B-Alert Web: an open-access tool for predicting the effects of sleep/wake schedules and caffeine consumption on neurobehavioral performance." *Sleep* (2016) 39: 2157–2159.

ROEHRS, T.A. ET AL. "Daytime sleepiness and antihistamines." *Sleep* (1984) 7: 137–141.

SAPER, C.B. AND BRADFORD B. LOWELL, B.B. "The hypothalamus." *Current Biology* (2014) 24: R1111–1116.

SAPER, C.B. ET AL. "Hypothalamic regulation of sleep and circadian rhythms." *Nature* (2005) 437: 1257–1263.

SIEGEL, J.M. "The neurotransmitters of sleep." *Journal of Clinical Psychiatry* (2004) suppl. 16: 4–7.

SILBER, M.H. "Staging sleep." *Sleep Medicine Clinics* (2012)7: 487–496.

STEIGER, A. "Sleep and endocrinology." *Journal of Internal Medicine* (2003) 254: 13–22.

STEPHAN, F.K. AND ZUCKER, I. "Circadian rhythms in drinking behavior and locomotor motility of rats are eliminated by hypothalamic lesions." *Proceedings of the National Academy of Science* (1972) 69: 1583–1586.

TAMISIER, R., OZAN TAN C., PEPIN J.-L., LEVY P. AND TAYLOR, J.A. "Blood pressure increases in OSA due to maintained neurovascular sympathetic transduction: impact of CPAP." *Sleep* (2015) 38: 1973–1980.

TANG, N.K.Y. ET AL. "Nonpharmacological treatments of insomnia for long-term painful conditions: a systematic review and meta-analysis of patient-reported outcomes in randomized controlled trials." *Sleep* (2015) 38: 1751–1764.

URSIN, R. "Serotonin and sleep." *Sleep Medicine Reviews* (2002) 6: 55–67.

WU ZHAO H., STEVENS, R.G., TENNEN, H., NORTH, C.S., JAMES J. GRADY, J.J. AND HOLZER, C. "Sleep quality among low-income young women in southeast Texas predicts changes in perceived stress through Hurricane Ike." *Sleep* (2015) 38: 1121–1128.

ZHANG, B. AND WING, Y.-K. "Sex differences in insomnia: a meta-analysis." *Sleep* (2006) 29: 85–93.

INDEX

A
acetylcholine 25, 26, 35, 36, 37, 100, 103
adenosine 36, 37, 38
adolescence 30, 40, 62, 92–3, 131, 156
adrenaline 25, 35
adrenocorticotropic hormone (ACTH) 104–5
adults *see* older people; young adults
advanced sleep phase syndrome 84–5
air travel 55, 63, 82–3, 167
  jet lag 82–3
alcohol 58, 82, 83, 116, 129, 130, 133, 136, 142–3
alpha waves 12–13, 16, 17, 58
Alzheimer's disease *see* dementia
amnesia 167
animals 10, 102
  dreaming 44, 50–1
  evolution of sleep 47–9
  sleep deprivation 44, 47, 54–5, 66–7, 68
antihistamines 168
Aristotle 44, 50
armodafinil 83, 129
Aschoff, Jurgen 74
Aserinsky, Eugene 14
astronauts 63
atonia 39, 50, 51, 133
auditory arousal threshold 11
Australian bearded dragon 47
autonomic nervous system (ANS) 24, 25, 27, 58

B
barbiturates 13, 158, 160, 163
bees 47, 74
benzodiazepines 13, 121, 123, 131, 133, 158–60, 167
Berger, Hans 12
Berger waves *see* alpha waves
beta waves 12, 13
birds 48, 51
body clock *see* circadian rhythms
brain 25, 27, 31, 32
  anatomy 34
  sleep deprivation 59, 64–5
  see also neurotransmitters
brain imaging studies 10, 93, 149
brain stem 26, 32, 33, 34, 36
brain waves 12, 13, 14
Bremer, Frederic 32
bullfrogs 46
Buysse, Daniel 30

C
caffeine 57, 63, 83, 140, 154, 156
Cartwright, Rosalind 40
cataplexy 124, 125, 129, 133
Caton, Richard 12
cats 32, 44, 50, 51
cell phones/tablets 90, 156
central sleep apnea 110–11, 113, 119, 153
cetaceans 49
children 20, 28, 30, 40, 41, 62, 88–91, 131, 152, 153, 156
chronic kidney disease (CKD) 140–1
circadian rhythms (body clock) 15, 31, 34, 47, 65, 74–7
  disorders 82–5, 153
  models 80–1
clonazepam 123, 131, 133
cockroaches 44–5, 47
cognitive behavioral therapy (CBT) 94, 139, 143, 153, 154, 157, 167
consciousness 10–11, 15, 44, 50
cortisol 74, 104–5, 107
cytokines 64, 66, 67, 141

D
death/mortality 28, 44, 45, 55, 62, 119, 164
deer 51
delayed sleep phase syndrome 79, 84, 85
Delorme, J.F. 50
delta sleep *see* slow-wave sleep
delta waves 13, 18, 19, 58, 96
Dement, William 56
dementia/Alzheimer's disease 65, 85, 94, 96, 97, 140
depression 20, 26, 27, 28, 68–9, 136–9, 149
diabetes 28, 30, 64, 107, 110, 115, 147
diazepam 49, 158
Dick, Philip K. 50
Dickens, Charles 111
diminished consciousness 10–11
dogs 44, 50, 54–5
dolphins 48, 49
dopamine 35, 121
Dorsey, Jack 54
doxepin 162–3, 164
dreaming 26, 40–1, 132–3
  animals 44, 50–1
  driving 116
  insomnia and 146–7
  medication and 129, 165, 166, 167, 168
  narcolepsy and 126
  sleep deprivation and 58, 63, 65
ducks 48
duration of sleep 28–31

E
Edison, Thomas 28, 54
Einstein, Albert 28
elderly people *see* older people
electrical phenomena 32–3
electroencephalogram (EEG) 12–13, 22–3, 26, 27, 48, 112
  stages of sleep 14–19
electromyogram (EMG) 12, 16, 19, 22–3, 24, 26, 27, 112, 113
electrooculogram (EOG) 12, 14, 19, 22–3, 26, 112
elephants 48
Eliot, George 124
encephalitis lethargica 32
ENTRAIN 29
epinephrine 25, 35
erections 27

F
Feinberg, Irwin 93
fight-or-flight response 35, 149
first night effect 49, 51
fish 45, 47
flu epidemic 1918 32
follicle-stimulating hormone 101, 107
Forel, Auguste 74
Franklin, Benjamin 81
free running disorder 79, 85
Freud, Sigmund 40, 50
fruit flies 47, 80
function of sleep 31, 70

G
GABA (gamma-aminobutyric acid) 36, 37, 38, 39, 59, 142
GABA agonists 83, 131, 160, 163
  medication and 158, 159, 160, 162, 168
Galvani, Luigi 12
genetic factors 28, 57, 80, 83
glial cells 31, 37
glutamate 35, 36, 37
glymphatic system 31, 64, 70
grehlin 64, 107
growth hormone (GH) 90, 96, 100, 102–3

H
hallucinations 55, 57, 58, 124, 125, 126
Hartmann, Ernest 40
headaches 79, 103, 134
herbal preparations 168–9
hippocampus 40, 50–1
histamine 36, 37, 126, 168
Hobson, Alan 38
homeostatic drive 31, 81, 96
homeostatic mechanism 31, 44, 46, 138
hormone replacement therapy 94, 110
Houasse, Michel Ange 143
Housman, A.E. 119
hyperarousal 148, 149
hypersomnia 110
hypertension 28, 30, 64, 110, 147, 149
hyperthyroidism 106, 140
hypoglossal nerve stimulation 118
hypothalamus 32, 33, 35, 36, 64, 100, 101, 142

I
idiopathic hypersomnia 110, 125, 126–9
immune system 64, 67, 70, 126, 141
impotence 27
infants 14, 20, 28, 62, 88–91, 134
insects 44–5, 47, 74, 80
insomnia 11, 54, 78, 83, 84, 93, 146–53
  depression 136, 149
  herbal preparations 168–9
  medication 149, 153, 158–67
  over-the-counter aids 163, 168–9
therapies 154–7
irregular sleep-wake rhythm 85

J
jet lag 82–3
Jouvet, Michel 50

K
K complexes 15, 17, 18, 19, 26, 89–90, 135
Kales, Anthony 16
Kleitman, Nathaniel 14, 15, 33, 56

L
leopards 51
leptin 64, 107, 111
Lindbergh, Charles 55
lions 51
locus coeruleus 34, 35, 37, 38, 50, 127
long sleepers 28
Loomis, Alfred Lee 14, 15
luteinizing hormone (LH) 101, 106–7

M
McCarley, Robert 38
macrophages 66, 141
Magoun, Horace Winchell 32
medication 11, 26, 58–9, 119, 123, 128–9, 131, 139, 140
antidepressants 66, 68, 139, 149, 153, 160, 162, 163
concerns about 62, 164–7
hypnotics 160, 164–5, 166
insomnia 149, 153, 158–67
off-label medicines 143, 163
melatonin 76, 77, 78–9, 83, 84, 85, 156, 162
Melville, Herman 124
memory 74, 96–7, 115, 126, 168
amnesia 167
dreaming and 40, 41
parasomnias and 130, 131, 132, 167
sleep deprivation and 57, 58, 59, 63, 66, 69, 105
menstrual cycle 93, 94
methysergide 103
mice 67, 102
Millais, John Everett 131
monkeys 51
Moruzzi, Giuseppe 32
Multiple Sleep Latency Test (MSLT) 56, 59, 63, 96, 128

N
Nagel, Thomas 10
Napoleon 28
naps 28, 56, 63, 68, 83, 125, 126, 128, 131, 152, 154
children 89
older people 96, 97
REM–NREM cycle 20
narcolepsy 95, 103, 124
narcolepsy–cataplexy 44, 124–6, 133
treatment 128–9
national differences 29, 60–1
neuroimaging studies 10, 58, 59
neurotransmitters 55, 100
effects on sleep 36–9
pathways 34–5
terminology 37
nicotine 57, 140, 154
night terrors 131
night-time gastroesophageal reflux (GER) 140
nightmares 41, 132–3
non-rapid eye movement (NREM) sleep 14, 15, 16, 36
disorders 41
dreams 40, 41
physiology of 24–5
REM–NREM cycle 20–1, 33, 38–9
norepinephrine 25, 34, 35, 36, 37, 55, 78, 100

O
obesity 30, 110, 116
obesity hypoventilation syndrome 111
obstructive sleep apnea (OSA) 27, 65, 90, 94, 95, 107, 110–19, 134, 141
older people 20, 30, 57, 62, 70, 83, 85, 96–7, 102, 120, 146, 162, 168
orexin/hypocretins 36, 37, 38, 39, 64, 115, 126, 127, 163

P
parasomnias 41, 110, 130–3, 164, 167
parasympathetic nervous system 24, 25, 26
penile tumescence 27
peri-locus coeruleus 26, 50
periodic limb movement disorder (PLMD) 22, 94, 95, 122–3, 140, 149, 153
pigeons 48
pituitary gland 78, 100–1, 102, 103, 104, 105, 106, 107
Poe, Edgar Allen 124
polysomnography (PSG) 22, 49, 112–15, 120, 122, 127, 150–1
positive airway pressure (PAP) 116–17, 118, 119
positron emission tomography (PET) imaging 24, 25, 27
post-traumatic stress disorder (PTSD) 41, 139
prazosin 133, 139
Psychomotor Vigilance Task (PVT) 59, 63, 70

R
ramelteon 83, 162, 163, 164
rapid eye movement (REM) sleep 14–15, 16, 19, 22, 36
animals 47, 48, 50–1
depression 136–7
deprivation 63, 66–7, 68
dreams 40, 41
induction of 26
physiology of 26–7
REM behavior disorder (RBD) 41, 50, 133
REM latency 20
REM–NREM cycle 20–1, 33, 38–9
rats 50–1, 55, 66
Rechtschaffen, Alan 16
Rechtschaffen-Kales criteria 16, 19
recording sleep 22–3
recovery sleep 10, 31, 47, 70–1
reptiles 47
respiration 24, 27, 91, 110–19, 124, 129, 166
restless legs syndrome (RLS) 120–1, 123, 131, 140, 141
reticular activating system (RAS) 32, 33, 34
rhythmic movement disorder 134
Rousseau, Henri 41

S
Saper, Clifford 39
sea otters 51
seizures 58, 90, 134, 135
serotonin 34, 35, 36, 37, 55, 78, 94, 100, 103
shift work 65, 83
short sleepers 28, 64, 69
sleep debt 31
sleep deprivation 15, 28, 31
acute 56–9
animals 44, 47, 54–5, 66–7, 68
benefits 68–9, 139
driving and 58, 63, 65
effects 54–5
pain resistance 67
partial 60–5
physiological effects 58–9, 64–5
REM sleep 63, 66–7, 68
selective 66–7
sleep disorders 41, 82–5
children 90
sleep need 31, 62
sleep paralysis 124, 126, 133
sleep positions 11
sleep spindles 17, 18, 19, 26, 89, 150
sleep–wake cycle 15, 37, 74, 80–1, 100
sleeping pills *see* medication
sleepwalking 41, 90, 130–1
slow-wave sleep (N3 sleep) 19, 20, 24, 47, 49, 59, 93, 96, 130, 140, 142, 168
deprivation 67, 69
hormone secretion 102, 103, 106
snoring 111, 119
sodium oxybate 58–9, 102, 129
space travel 63
stages of sleep 14–19, 48, 89–90
across the night 20–1
N1/stage 1 17, 20, 22, 27
N2/stage 2 13, 18, 20, 24, 26, 130, 150
N3 sleep/stages 3 and 4 *see* slow-wave sleep
recording sleep 22–3
Stickgold, Robert 40
suprachiasmatic nucleus (SCN) 34, 74–7, 78, 82
suvorexant 163, 164
sympathetic nervous system 24, 25, 77, 142, 149

T
thalamus 32, 33, 59
Thatcher, Margaret 28
theta waves 13, 26, 58
thyroid-stimulating hormone (TSH) 101, 106
tiagabine 59
trazodone 143, 160, 163
Tripp, Peter 56, 57, 58
twins 28, 126

V
valerian 168–9
Vogel, Gerald 68
von Economo, Constantin 32

W
Wever, Rutger 74
whales 48, 49

Y
young adults 20, 21, 30, 62, 63, 66, 70, 79, 88, 94–5, 103

Y
zebra finch 48
zebrafish 45, 47
zeitgebers 76, 80, 81
zolpidem 160, 161, 163

ACKNOWLEDGMENTS

AUTHOR ACKNOWLEDGMENTS

I want to thank Dr. Russell Rosenberg for generously supplying a physiologic recording sample. A number of colleagues have read materials from this book and given me comments and suggestions. I greatly appreciate the thoughtfulness of Drs. Orfeu Buxton, Daniel J. Buysse, Rosalind D. Cartwright, Karl Doghramji, David F. Dinges, Meir H. Kryger, Adrian R. Morrison, Russell Rosenberg, Michael A. Schwartz and James K. Walsh. Any errors which remain are my own.

PICTURE CREDITS

The publisher would like to thank the following for permission to reproduce copyright materials.

11: Ted Spagna/Science Photo Library.

15 left: Library of Congress, Washington DC.

15 right: Special Collections Research Center, University of Chicago Library.

17 top (EOG eye channels): Dr. Russell Rosenberg.

21: Source: Kathryn Lovell, PhD and Christine Liszewski, MD. "Normal Sleep Patterns and Sleep Disorders." http://learn.chm.msu.edu/NeuroEd/neurobiology_disease/content/otheresources/sleepdisorders.pdf

23 right: Omikron/Science Photo Library.

24–25: Hank Morgan/Science Photo Library.

28 left: Library of Congress, Washington DC.

28 right: Georgios Kollidas/Shutterstock.

29: Source: Olivia J. Walch, Amy Cochran, and Daniel B. Forger, "A global quantification of 'normal' sleep schedules using smartphone data" in *Science Advances* (6 May 2016) Vol. 2, no. 5, e1501705. DOI: 10.1126/sciadv.1501705. This work is licensed under CC BY-NC (http://creativecommons.org/licenses/by-nc/4.0/).

30: Source: Bureau of Labor Statistics, American Time Use Survey 2015.

38: Republished with permission of the American Physiological Society, from Ritchie E. Brown, Radhika Basheer, James T. McKenna, Robert E. Strecker, and Robert W. McCarley, "Control of Sleep and Wakefulness" in *Physiological Reviews* (July 2012) 92 (3), 1087–1187; permission conveyed through Copyright Clearance Center, Inc.

39: Republished with permission of the American Academy of Sleep Medicine, from Clifford B. Saper, Patrick M. Fuller, Nigel P. Pedersen, Jun Lu, and Thomas E. Scammell, "Sleep State Switching" in *Neuron* (Dec 2010) 68 (6), 1023–1042; permission conveyed through Copyright Clearance Center, Inc.

41: Buyenlarge/Hulton Fine Art Collection/Getty Images.

45 top: seeyou/Shutterstock.

45 bottom: Grigorev Mikhail/Shutterstock.

46: Kenneth H. Thomas/Science Photo Library.

47: hddigital/Shutterstock.

48 centre: xpixel/Shutterstock.

48 bottom: Adapted from P.S. Low, Sylvan S. Shank, T.J. Sejnowski, and D. Margoliash, "Mammalian-like features of sleep structure in zebra finches" in PNAS (2008) 105: 9081–9086. Copyright (2008) National Academy of Sciences, U.S.A.

51 top: Smokedsalmon/Shutterstock.

51 bottom: Achim Baque/Shutterstock.

54: Everett Historical/Shutterstock.

55: Bettmann/Getty Images.

57: Ted Russell /The LIFE Images Collection/Getty Images.

59: Republished with permission of the American Academy of Sleep Medicine, from Ning Ma, David F. Dinges, Mathias Basner, and Hengyi Rao "How Acute Total Sleep Loss Affects the Attending Brain: A Meta-Analysis of Neuroimaging Studies" in *Sleep* (2015) 38 (2): 233-240; permission conveyed through Copyright Clearance Center, Inc.

60: Yoshikazu Tsuno/AFP/Getty Images.

61: Sources: OECD, Australian Bureau of Statistics, Statistics New Zealand, United States Bureau of Labor Statistics, Statistics Canada, Japanese Time Use Survey.

http://www.huffingtonpost.com/2013/08/24/average-daily-nightly-sleep-country-world_n_3805886.html

62: Source: National Sleep Foundation (2015).

http://www.prevention.com/health/sleep-energy/are-you-getting-enough-sleep-based-your-age?adbpr=25092348&adbid=625093599311454208&adbpl=tw&cid=socHE_20150726_49647936

age?adbpr=25092348&adbid=625093599311454208&adbpl=tw&cid=socHE_20150726_49647936

63: Vadim Sadovski/Shutterstock.

64: C.J. Guerin, PhD, MRC Toxicology Unit/Science Photo Library.

66: Nancy Kedersha/Science Photo Library.

67: Republished with permission of the American Academy of Sleep Medicine, from Timothy Roehrs, Maren Hyde, Brandi Blaisdell, Mark Greenwald, and Thomas Roth "Sleep Loss and REM Sleep Loss are Hyperalgesic" in *Sleep* (2006) 29 (2): 145–51; permission conveyed through Copyright Clearance Center, Inc.

69: Republished with permission of the American Academy of Sleep Medicine, from K. Porcheret, E.A. Holmes, G.M. Goodwin, R.G. Foster, and K. Wulff, "Psychological Effect of an Analogue Traumatic Event Reduced by Sleep Deprivation" in *Sleep* (2015) 38 (7): 1017–25; permission conveyed through Copyright Clearance Center, Inc.

71: Republished with permission of the American Academy of Sleep Medicine, from: M.A. Carskadon and W.C. Dement, "Sleep loss in elderly volunteers" in *Sleep* (1985) 8:207–221; permission conveyed through Copyright Clearance Center, Inc.

79: Republished with permission of the Oxford University Press Journals, from: George M. Vaughan, Russell W. Pelham, Shiu F. Pang, Leo L. Loughlin, Kenneth M. Wilson, Kenneth L. Sandock, Mary K. Vaughan, Stephen H. Koslow, and Russel J. Reiter "Nocturnal Elevation of Plasma Melatonin and Urinary 5-Hydroxyindoleacetic Acid in Young Men: Attempts at Modification by Brief Changes in Environmental Lighting and Sleep and by Autonomic Drug" in *The Journal of Clinical Endocrinology and Metabolism* (2016) 42 (4): 752–764; permission conveyed through Copyright Clearance Center, Inc.

81 bottom: Library of Congress, Washington DC.

83: yuttana Contributor Studio/Shutterstock.

89: Republished with permission of the American Academy of Sleep Medicine, from: M.M. Ohayon, M.A. Carskadon, C. Guilleminault, and M.V. Vitiello, "Meta-Analysis of Quantitative Sleep Parameters From Childhood to Old Age in Healthy Individuals: Developing Normative Sleep Values Across the Human Lifespan" in *Sleep* (2004) 27 (7): 1255–73; permission conveyed through Copyright Clearance Center, Inc.

90: Anna Goroshnikova/Shutterstock.

91: Musée d'Orsay, Paris.

92: Michael Whitehead.

93: Republished with permission of the American Academy of Sleep Medicine, from: Irwin Feinberg, Evan de Bie, Nicole M. Davis, and Ian G. Campbell, "Topographic Differences in the Adolescent Maturation of the Slow Wave EEG during NREM Sleep" in *Sleep* (2011) 34 (3): 325–333; permission conveyed through Copyright Clearance Center, Inc.

95 top: Kate Zubal/Shutterstock.

95 bottom: JC-PROD/Shutterstock.

97: Source: National Sleep Foundation, 2003 Sleep in America poll.

103 bottom: Steve Gschmeissner/Science Photo Library.

121: JL-Pfeifer/Shutterstock.

122: Phanie/Alamy.

125: Hoang Dinh Nam/AFP/Getty Images.

129: Kinga/Shutterstock.

131: Private Collection/Bridgeman Images.

132: CREATISTA/Shutterstock.

135: Republished by permission of Oxford University Press, from: Federica Provini, Giuseppe Plazzi, Paolo Tinuper, Stefano Vandi, Elio Lugaresi, Pasquale Montagna, "Nocturnal frontal lobe epilepsy: A clinical and polygraphic overview of 100 consecutive cases" in *Brain* (1999) 122 (6): 1017–1031; permission conveyed through Copyright Clearance Center, Inc.

141: Science Photo Library.

143: Heritage Images/Hulton Fine Art Collection/Getty Images.

147: Kimmo Metsaranta/Folia Images/Getty Images.

148: Republished with permission of the American Academy of Sleep Medicine, from: M. Desseilles, T. Dang-Vu, M. Schabus, V. Sterpenich, P. Maquet, and S. Schwartz, "Neuroimaging Insights into the Pathophysiology of Sleep Disorders" in *Sleep* (2008) 31(6): 777–794; permission conveyed through Copyright Clearance Center, Inc.

151: Aubert/BSIP/Science Photo Library.

152: J. Parsons/Getty Images.

155: Lucas Pham/EyeEm/Getty Images.

156: Peter Snaterse/Shutterstock.

161: Source: Thomas Roth, Timothy Roehrs, and Gerald Vogel, "Zolpidem in the treatment of transient insomnia: a double-blind randomized comparison with placebo" in *Sleep* (1995) 18(4): 246–251.

162: Source: Centers for Disease Control and Prevention/National Center for Health Statistics, National Health and Nutrition Survey. From Chong Y, Fryar CD, Gu Qu "Prescription Sleep Aid Use Among Adults: United States 2005–2010" in NCHS Data Brief No 127. Hyattsville, MD: National Center for Health Statistics 2013.

165: Source: T. A. Betts and Janice Birtle: "Effect of two hypnotic drugs on actual driving performance next morning" in *British Medical Journal*, Volume 285, 25 September 1982, page 852.

166: Eric Savage/Getty Images.

169: Klarsichtstudio/Shutterstock.

ILLUSTRATIONS AND GRAPHICS

Note: In some cases, raw data from physiologic recordings such as EEG, EMG, and EOG have been modified to remove artifacts or otherwise for purposes of clarity.

Ivan Hissey: 123.

Louis Mackay: 23, 33, 34, 35, 37 top right, 65, 77, 100, 105, 106, 107, 111, 113, 117, 118, 127, 148, 159.

Lisa McCormick: 21, 30, 38, 61, 62, 67, 69, 81 top, 85, 88, 89, 97, 100, 137, 138, 161, 162, 165

Richard Palmer: 49.

Richard Peters: 82.

Nick Rowland: 29.

John Woodcock: 13, 14, 16–19, 27, 37 left, 39, 71, 75, 79, 84, 93, 103, 104, 112, 114.